SpringerBriefs in Computer Science

Series Editors
Stan Zdonik
Peng Ning
Shashi Shekhar
Jonathan Katz
Xindong Wu
Lakhmi C. Jain
David Padua
Xuemin (Sherman) Shen
Borko Furht
V.S. Subrahmanian
Martial Hebert
Katsushi Ikeuchi
Bruno Siciliano

For further volumes:
http://www.springer.com/series/10028

Subodha Gunawardena • Weihua Zhuang

Modeling and Analysis of Voice and Data in Cognitive Radio Networks

Springer

Subodha Gunawardena
Department of Electrical
  and Computer Engineering
University of Waterloo
Waterloo, ON, Canada

Weihua Zhuang
Department of Electrical
  and Computer Engineering
University of Waterloo
Waterloo, ON, Canada

ISSN 2191-5768   ISSN 2191-5776 (electronic)
ISBN 978-3-319-04644-0   ISBN 978-3-319-04645-7 (eBook)
DOI 10.1007/978-3-319-04645-7
Springer Cham Heidelberg New York Dordrecht London

Library of Congress Control Number: 2014931761

Printed on acid-free paper

Springer is part of Springer Science+Business Media (www.springer.com)

Preface

The growing interest towards wireless communication services over the recent years has increased the demand for radio spectrum. Inefficient spectrum management together with the scarcity of the radio spectrum is a limiting factor for the development of modern wireless networks. As a solution, the idea of cognitive radio networks (CRNs) is introduced to use licensed spectrum bands for the benefit of the unlicensed secondary users. However, the preemptive priority of the licensed users results in random resource availabilities at the secondary networks, which makes the quality-of-service (QoS) support challenging. With the increasing demand for elastic/interactive data services (internet based services) and wireless multimedia services, QoS support becomes essential for CRNs. This brief investigates the voice and elastic/interactive data service support over CRNs in terms of their delay requirements. The packet level delay requirements of the voice service and session level delay requirements of the elastic/interactive data services are considered. In particular, constant-rate and on-off voice traffic capacities are analyzed over CRNs with centralized and distributed network coordination. Some generic channel access schemes are considered as the coordination mechanisms, and call admission control algorithms are developed for non-fully-connected CRNs. Advantage of supporting voice traffic flows with different delay requirements in the same network is also discussed. The mean response time of the elastic data traffic over a centralized CRN is studied, considering the shortest processor time with and without preemption and shortest remaining processor time service disciplines, in comparison with the processor sharing service discipline. Effects of the traffic load at the base station and file length (service time requirement) distribution on the mean response time are discussed. Finally, the relationship between the mean response times of interactive and elastic data traffic is presented.

Waterloo, ON, Canada   Subodha Gunawardena
Waterloo, ON, Canada   Weihua Zhuang
October 2013

Contents

Chapter 1
Introduction

1.1 Dynamic Spectrum Access

The extensive growth of wireless networks over the recent years has increased the demand for the radio spectrum to a great extent. Static spectrum allocation regardless of its spatiotemporal usage has led to scarcity of the spectrum. As a result, fulfilling the spectrum requirements of emerging wireless applications and technologies is a challenging task. However, it is shown by various research groups [1,2] that the spectrum utilizations in different locations are well below 25% in most of the spectrum bands, and the spectrum scarcity is mostly due to the inefficient spectrum management. As a solution to this problem, dynamic spectrum access methods have been introduced. They can be categorized based on the regulatory status, namely, the dynamic licensing and dynamic sharing, as illustrated in Fig. 1.1.

In dynamic licensing, the authority of a given spectrum band is exclusively assigned to a particular operator for a given period of time. Therefore, the network that can generate the maximum profit for the given time period can obtain the right to access spectrum. It is more flexible than the static licensing used by today's spectrum regulators. However, both licensing methods are based on an exclusive-use model, which is limited in the adaptation speed. In dynamic spectrum sharing, more than one network operator is given the authority to use the same spectrum band (co-existence). It is categorized into horizontal and vertical spectrum sharing. Horizontal sharing is the co-existence of networks with equal regulatory status, whereas the vertical sharing is the co-existence of networks with different regulatory status. A special type of vertical spectrum sharing concept was proposed in late 1990s to share the licensed spectrum bands with the unlicensed networks [4,5], and these unlicensed networks are named as cognitive radio networks (CRNs).

S. Gunawardena and W. Zhuang, *Modeling and Analysis of Voice and Data in Cognitive Radio Networks*, SpringerBriefs in Computer Science, DOI 10.1007/978-3-319-04645-7__1, © The Author(s) 2014

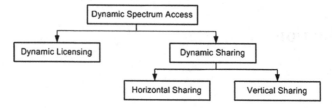

Fig. 1.1 Categorization of dynamic spectrum access [3]

1.2 The Concept of CRNs

The concept of cognitive radio networks is to make use of underutilized portions of the spectrum by operating unlicensed (secondary) networks over the licensed (primary) spectrum bands. The basic functionality of a CRN is to use the under-used/unused portions of the licensed spectrum for the benefit of the unlicensed users, without causing harmful interference to the licensed users [4–6]. The licensed users are called primary users (PUs) and the unlicensed users are called secondary users (SUs). The concept of cognitive radio has been well accepted within the wireless communications research community. With the rapid development of smart handsets, realizing the CRNs become more realistic, and it is predicted to be an important component of the wireless communication networks.

The CRNs are categorized into two types, namely, the underlay and overlay networks. In the underlay CRNs, the SUs are allowed to access the licensed channels simultaneously with the PUs. Therefore, the spectrum bands are always available for the SUs. However, it is required to control the transmit power of the SUs to protect the PUs from interference. Due to this phenomena, in general, the underlay CRNs are smaller in size (coverage area) [7]. In overlay networks, the SUs are allowed to access licensed channels only when the PUs are not using them. Therefore, the spectrum bands are not always available for the secondary network, and the CRN concept is based on the secondary users' awareness of the primary user dynamics. SUs are required to sense the spectrum bands to identify the spectrum opportunities. The IEEE 802.22 [8] is an example for overlay networking standard. In this brief, we focus on the overlay CRNs, and from this point onward, we use the term cognitive radio network to denote an overlay CRN.

1.3 Challenges

The main objective of the secondary network is to exploit the maximum amount of spectrum opportunities without harmfully interfering the PUs. The spectrum availability exclusively depends on the behavior of the PUs. As the activities of the PUs are not known to the secondary network, the spectrum sensing holds the key in enabling the CRNs. The spectrum sensing consists of obtaining the measurements, analyzing the acquired data, and decision making procedures. The faster the sensing (the shorter the sensing duration), the longer the channel time available for the secondary network. On the other hand, the shorter the sensing duration, the lower the accuracy of sensing results, leading to interference with the primary network. The sensing errors are inevitable in practical networking scenarios. The SUs are required to keep the sensing errors under a certain threshold in such a way that interference to the PUs can be controlled within the acceptable limits. Therefore, choosing the appropriate sensing mechanism to optimize the network performance is a challenging task.

Even if we assume the existence of a perfect spectrum sensing mechanism, the channel availability for the SUs are exclusively dependent up on the behavior of the PUs. Therefore, the node coordination in a CRN is more complicated than that of a conventional wireless network. Further, due to the random nature of the PU activities, the channel time available for the SUs is random. Based on the spacial distribution of the SUs, they may have different views about the PU activities. If all the SUs experience the same spectrum availability, it is denoted as a homogeneous network, and a heterogeneous network otherwise. Further, based on the availability of infrastructure, the CRNs can be categorized into infrastructure based and infrastructure-less CRNs, as illustrated in Fig. 1.2.

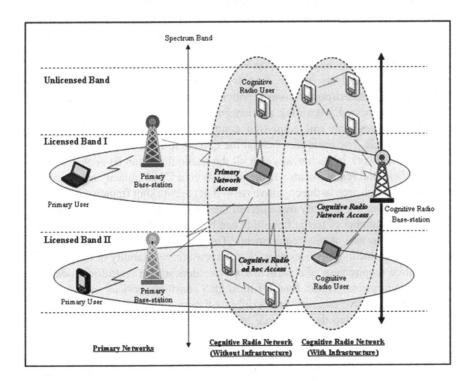

Fig. 1.2 Categorization of CRNs based on the availability of infrastructure [9]

The infrastructure-less (ad hoc) nature of a network makes it more flexible in adapting to its surroundings, which is more desirable in the context of CRNs. However, the spectrum non-homogeneity and infrastructure-less nature of CRNs can make the node coordination and information packet transmission extremely challenging. Further, the randomness of the spectrum resource availability of the secondary network makes it challenging task to estimate the resource availability of the network. Therefore, the initial research studies on CRNs address only the best effort type of services without strict service requirements. The ongoing research studies on CRNs can be broadly categorized as: (1) the realization of CRNs by better spectrum sensing, channel access, interference avoidance, routing, and topology management techniques [10–17]; (2) the performance aspects of CRNs such as system capacity, fairness, throughput, and revenue [18–37].

1.4   Quality-of-Service (QoS) Support

The rapid growth of Internet based interactive services and wireless multimedia services over the conventional wireless networks demands the cognitive radio research community to work on QoS support over the CRNs. These services demand for certain quality satisfaction by the underlying network. Guaranteeing the service quality requires an estimation of the available amount of resources and an efficient coordination among the SUs, which are equally challenging due to the time varying nature of the PU activities. Over the past few years, significant efforts have been made towards enabling the QoS support via efficient channel access control [18–24] and performance analysis of QoS sensitive services [25–37] over CRNs. In the studies, channel access/power control policies, cross-layer designs, and capacity analysis are discussed. Providing service guarantees requires the knowledge of network resources, and call admission control (CAC) to maintain the service satisfaction levels. The basic idea behind the CAC is to limit the number of traffic flows in the network in such a way that all the traffic flows can be served without violating their QoS requirements. As the foundation of CAC, capacity analysis is an important research area in the cognitive radio (CR) networking framework.

Different from the capacity definition of the physical layer[1] [38], the capacity is defined as the number of traffic flows that can be supported by the system, without violating their service quality requirements. The capacity of a CRN depends on the resource requirement of services as well as the resource availability of the network. The resource requirement of a particular service depends on its traffic model and its QoS requirements. Traffic flows of most QoS sensitive services such as voice, video, and interactive data services exhibit a time varying nature, which makes it difficult to determine their resource requirements. Furthermore, estimating the

---

[1]The physical layer capacity only depends on the channel bandwidth, availability, and its condition, and it is given in terms of either bits/channel use or bits/second.

resource availability of CRNs becomes complicated with the random nature of the spectrum availabilities, which depends on the PU activities and the location of the SUs. The service disciplines of the base stations (or central controllers) in centralized networks and coordination mechanisms in distributed networks can make a significant impact on the resource utilization (the fraction of the exploited amount of resources from what is available) of the SUs. The randomness of the network resource availability, the time varying nature of resource requirements, and stringent QoS requirements of the wireless services make the capacity analysis and admission important and control challenging.

1.5 Motivations

As two of the common QoS sensitive services, voice and (elastic/interactive) data services have gained an immense popularity. The voice service falls under the real-time traffic category, whereas the elastic/interactive data services falls under the non-real-time traffic category. However, both the services have service requirements in terms of delay. The voice service has a packet level service requirement in terms of end-to-end packet delay, and elastic/interactive data services have a session level service requirement in terms of file transfer time (also known as the response time). These services are among the most popular services over the conventional wireless networks, and their QoS support is extensively studied over the past two decades. As an emerging next generation networking concept, it is necessary to support these services over CRNs. Although there have been numerous works on voice service support over CRNs [25–31, 37], much attention has not been focused on the legacy channel access schemes in supporting voice service, which can be used to provide useful benchmarks to the CR networking research community. Further, it is interesting to study the aspect of supporting voice calls with different delay requirements, which gives the SUs an option to choose the required service quality depending on the service cost. Studies on elastic/interactive data traffic services such as web browsing are carried out over the conventional wireless networks [39, 40]. There are no significant service interruptions for the data services operating over these networks when compared with that of the CRNs. The service interruptions due to the presence of PUs increase the response time of SU elastic/interactive data traffic flows. Different service disciplines can be used to provide certain priorities to different users based on their requested file lengths. So far, only a handful of research efforts are devoted to the elastic/interactive data services over CRNs.

Considering the aforementioned motivational facts, we will take the following steps to present the voice capacity analysis and elastic/interactive data service support over CRNs [41].

1.6 Contributions

In Chap. 2, we present the voice capacity of centralized and distributed overlay CRNs. The capacity is given in terms of the number of simultaneous voice calls that can be supported by the system. Specifically speaking, we analyze the voice capacity of single-channel fully-connected CRNs considering the packet delay as the QoS measure. Under this, we study: (1) The on-off and constant-rate voice capacity of a centralized CRN with first-come first-serve (FCFS) service discipline at the base station (BS) [42]; (2) The constant-rate voice capacity of a single-channel distributed fully-connected CRN under different channel access schemes, and compare the performance of different schemes [43]. Considering the slot-ALOHA, round-robin, and random allocation schemes, we study the effect of channel rate (the number of voice packets that can be transmitted in a time-slot) and the channel unavailable duration, respectively, on the system capacity. Then, we develop a CAC procedure for a distributed non-fully-connected CRN with slot-ALOHA network coordination, assuming homogeneous single-hop voice traffic flows based on the capacity analysis results of the fully-connected network. Further, we develop two CAC algorithms for the voice traffic flows with different delay requirements [44].

In Chap. 3, we present the mean response time (MRT) of the elastic data service operating over a single-channel time-slotted centralized CRN with three service disciplines, namely, shortest processor time without preemption (SPTNP), shortest processor time with preemption (SPTWP), and shortest remaining processing time (SRPT), in comparison with the processor sharing (PS) service discipline [45, 46]. We compare all four service disciplines under different data traffic load conditions and Weibull distributed data file lengths with different tail properties. Further, we show that the variation of mean channel unavailable duration has a significant impact on the MRT, even when the probability of channel availability (long-term channel availability) remains unchanged. To the best of our knowledge, this is the first work that these service disciplines are considered with elastic data traffic in CRNs.

In Chap. 4, we present the MRT of the interactive data traffic service in which each session consists of multiple data file requests (service requests). We show that the request arrivals at the BS can be approximated by a Poisson process under specific channel availability conditions. Based on the Poisson approximation, we show that the MRT of interactive (multi-file) data sessions can be approximated by that of elastic (single-file) data traffic sessions with an equivalent session arrival rate.

1.7 Summary

In this chapter, we present a brief introduction to dynamic spectrum access techniques and the concept of cognitive radio networks. Further, we provide the basic categorizations of CRNs based on the spectrum access technique (underlay and overlay networks) and the SU view point of the available spectrum (homogeneous and non-homogeneous networks), respectively. Then, the importance and challenges in the QoS support over CRN are discussed. Finally, the motivation for the voice service support and the elastic/interactive data services support is discussed, and the important contributions of the subsequent chapters are summarized.

References

1. Belt, B., Khlopin, D., Seiffert, G.: Comments of the telecommunications industry association. Tech. Rep. 02-135, Telecommunications Industry Association (2003)
2. McHenry, M., Tenhula, P., McCloskey, D., Roberson, D., Hood, C.: Chicago spectrum occupancy measurements & analysis and a long-term studies proposal. In: Proc. ACM TAPAS (2006)
3. Hossain, E., Bhargava, V.: Cognitive wireless communication networks. Springer US (2007)
4. Mitola, J.: Cognitive radio: an integrated agent architecture for software defined radio. Dissertation, Doctor of Technology, Royal Institute of Technology, Sweden (2000)
5. Mitola, J., Jr., G.Q.M.: Cognitive radio: making software radios more personal. IEEE Pers. Commun. Mag. 6(4), 13–18 (1999)
6. Mitola, J.: Cognitive radio for flexible mobile multimedia communications. In: Proc. IEEE MOMUC, pp. 3–10 (1999)
7. Small cell forum. www.smallcellforum.org. Accessed 10 Sep. 2013.
8. Working group on wireless regional area networks, ieee 802.22. www.ieee802.org/22/. Accessed 10 Sep. 2013.
9. Broadband wireless networking lab, school of electrical and computer engineering, georgia institute of technology. www.ece.gatech.edu/research/labs/bwn/CR/. Accessed 10 Sep. 2013. (2013)
10. Hur, Y., Park, J., Woo, W., Lim, K., Lee, C.H., Kim, H.S., Laskar, J.: A wideband analog multi-resolution spectrum sensing (mrss) technique for cognitive radio (cr) systems. In: Proc. IEEE DySPAN, pp. 4090–4093 (2006)
11. Zhao, J., Zheng, H., Yang, G.: Distributed coordination in dynamic spectrum allocation networks. In: Proc. IEEE DySPAN, pp. 259–268 (2005)
12. Chen, T., Zhang, H., Maggio, G.M., Chlamtac, I.: Cogmesh: a cluster-based cognitive radio network. In: Proc. IEEE DySPAN, pp. 167–178 (2007)
13. Cai, Z., Lu, M., Wang, X.: Channel access-based self-organized clustering in ad hoc networks. IEEE Trans. Mobile Comput. 2(2), 102–113 (2003)
14. Choi, N., Patel, M., Venkatesan, S.: A full duplex multi-channel mac protocol for multi-hop cognitive radio networks. In: Proc. IEEE CROWNCOM, pp. 1–5 (2006)
15. Shu, T., Cui, S., Krunz, M.: Medium access control for multi-channel parallel transmission in cognitive radio networks. In: Proc. IEEE GLOBECOM, pp. 1–5 (2006)
16. Sankaranarayanan, S., Papadimitratos, P., Mishra, A.: A bandwidth sharing approach to improve licensed spectrum utilization. IEEE Commun. Mag. 43(12), S10–S14 (2005)
17. Timmers, M., Dejonghe, A., der Perre, L.V., Catthoor, F.: A distributed multichannel mac protocol for cognitive radio networks with primary user recognition. In: Proc. IEEE CROWNCOM, pp. 216–223 (2007)
18. Wang, L., Chen, A., Wei, D.: A cognitive mac protocol for qos provisioning in overlaying ad hoc networks. In: Proc. IEEE CCNC, pp. 1139–1143 (2007)
19. Su, H., Zhang, X.: Cross-layer based opportunistic mac protocols for qos provisioning over cognitive radio wireless networks. IEEE J. Select. Areas Commun. 26(1), 118–129 (2008)
20. Ishibashi, B., Bouabdallah, N., Boutaba, R.: Qos performance analysis of cognitive radio-based virtual wireless networks. In: Proc. IEEE INFOCOM (2008)

8 1 Introduction

21. Hong, X., Wang, C.X., Chen, H.H., Thompson, J.: Performance analysis of cognitive radio networks with average interference power constraints. In: Proc. IEEE ICC, pp. 3578–3582 (2008)
22. Zhang, M., Jiang, S., Wei, G., Wang, H.: Performance analysis of the cognitive radio network with a call level queue for secondary users. In: Proc. IEEE WiCOM (2009)
23. Zhang, M., Jiang, S., Wei, G., Wang, H.: Performance analysis of cognitive radio networks against secondary user's policies. In: Proc. IEEE ACIS (2009)
24. Ruan, L., Lau, V.K.N.: Power control and performance analysis of cognitive radio systems under dynamic spectrum activity and imperfect knowledge of system state. IEEE Trans. Wireless Commun. 8(9), 4616–4622 (2009)
25. Wong, W., Foh, H.: Analysis of cognitive radio spectrum access with finite user population. IEEE Commun. Lett. 13(5), 294–296 (2009)
26. Akin, S., Gursoy, M.C.: Performance analysis of cognitive radio systems under qos constraints and channel uncertainty. IEEE Trans. Wireless Commun. 10(9), 2883–2895 (2011)
27. Jang, Y.U.: Performance analysis of cognitive radio networks based on sensing and secondary-to-primary interference. IEEE Trans. Signal Proc. 59(11), 5663–5668 (2011)
28. Wang, P., Niyato, D., Jiang, H.: Voice service support over cognitive radio networks. In: Proc. IEEE ICC (2009)
29. Wang, P., Niyato, D., Jiang, H.: Voice service capacity analysis for cognitive radio networks. IEEE Trans. Wireless Commun. 59(4), 1779–1790 (2010)
30. Lee, H., Cho, D.H.: Voip capacity analysis in cognitive radio system. IEEE Commun. Lett. 13(6), 393–395 (2009)
31. Lee, H., Cho, D.H.: Capacity improvement and analysis of voip service in a cognitive radio system. IEEE Trans. Veh. Technol. 59(4), 1646–1651 (2010)
32. Cruz-Perez, F.A., Rivero-Angeles, M.E., Hernandez-Valdez, G., Castellanos-Lopez, S.L.: Joint call and packet level performance analysis of cac strategies for voip traffic in wireless. In: Proc. IEEE GLOBECOM (2011)
33. Qin, H., Cui, Y.: Cross-layer design of cognitive radio network for real time video streaming transmission. In: Proc. IEEE CCCM, pp. 376–379 (2009)
34. Mao, S., .Hu, D.: Streaming scalable videos over multi-hop cognitive radio networks. IEEE Trans. Wireless Commun. 9(11), 3501–3511 (2010)
35. Guan, B., He, Y.: Optimal resource allocation for video streaming over cognitive radio networks. In: Proc. IEEE MMSP, pp. 198–205 (2011)
36. Kartheek, M., Misra, R., Sharma, V.: Performance analysis of data and voice connections in a cognitive radio network. In: Proc. IEEE NCC (2011)
37. Koufos, K., Ruttik, K., Jantti, R.: Voice service in cognitive networks over the tv spectrum. IET Commun. Mag. 6(8), 991–1003 (2012)
38. Wyglinski, A.M., Nekovee, M., Hou, Y.T.: Cognitive Radio Communications and Networks: Principles and Practice. Elsevier (2009)
39. Song, W., Zhuang, W.: Multi-class resource management in a cellular/wlan integrated network. In: Proc. IEEE WCNC, pp. 3070–3075 (2007)
40. Song, W., Zhuang, W.: Resource allocation for conversational, streaming, and interactive services in cellular/wlan interworking. In: Proc. IEEE GLOBECOM, pp. 4785–4789 (2007)
41. Gunawardena, S.: Voice capacity and data response time in cognitive radio networks. Dissertation, Doctor of Philosophy, University of Waterloo, Canada (2013)
42. Gunawardena, S., Zhuang, W.: Voice capacity of cognitive radio networks. In: Proc. IEEE ICC (2010)
43. Gunawardena, S., Zhuang, W.: Voice capacity of cognitive radio networks for both centralized and distributed channel access control. In: Proc. IEEE GLOBECOM (2010)
44. Gunawardena, S., Zhuang, W.: Capacity analysis and call admission control in distributed cognitive radio networks. IEEE Trans. Wireless Commun. 10(9), 3110–3120 (2012)
45. Gunawardena, S., Zhuang, W.: Service response time of elastic data traffic in cognitive radio networks with spt service discipline. In: Proc. IEEE GLOBECOM (2012)
46. Gunawardena, S., Zhuang, W.: Service response time of elastic data traffic in cognitive radio networks. IEEE J. Select. Areas Commun. 31(3), 559–570 (2013)

Chapter 2
Voice Capacity Analysis

2.1 Introduction

The voice and video streaming are the most common real-time wireless services in the modern day wireless networks [1]. The demand for video streaming is growing rapidly, and it is predicted to be the most common wireless service in the future [2]. As an upcoming networking approach, it is necessary to support the voice and video services over CRNs.

The cellular and satellite networks have dedicated capacity allocation to voice traffic flows, and dedicated channel time is available for each flow. Therefore, the delay of a voice packet is a deterministic quantity, which depends on the propagation delay through the network. However, in the networks such as wireless local area networks (WLANs), voice traffic flows are sharing the network resources with other services such as video streaming and interactive data services. Therefore, fixed capacity allocation is not always feasible. Due to the random nature of the channel access, the packet delay is not deterministic. When the voice service is operating in such a network, the service quality requirement is given in terms of a stochastic delay requirement [3], such as $P(D > D_{max}) \leq \epsilon$, where D is the end-to-end delay of a voice packet, D_{max} and ϵ are the delay bound (or the maximum delay) and maximum delay bound violation probability allowed, respectively. Further, the service requirement is relaxed in some studies by giving the QoS requirement in terms of mean of the packet delay. Voice traffic sources are categorized into constant-rate and on-off traffic sources [4] based on the output of the voice coder. The voice service support is extensively studied over the conventional wireless networks, and as an upcoming networking approach, it is an important research area in the context of CRNs.

In an overlay CRN, the channel availability for the secondary network exclusively depends on the behavior of the PUs. Therefore, the amount of spectrum resources available for the secondary network is limited. In order to provide the required service levels using limited available resources in CRNs, capacity analysis and call admission control are essential. The voice service support over CRNs

S. Gunawardena and W. Zhuang, *Modeling and Analysis of Voice and Data in Cognitive Radio Networks*, SpringerBriefs in Computer Science, DOI 10.1007/978-3-319-04645-7__2, © The Author(s) 2014

has been studied in the literature with respect to the capacity analysis [5–8], call admission control [9, 10], and developing channel access schemes [5, 6, 11]. The voice capacity is defined as the maximum number of voice calls that can be supported by the network satisfying their service requirements. In [7, 8, 11], the network under consideration is a centralized CRN, and the service quality is measured in terms of the packet dropping probability due to the buffer overflow at the base station [7, 8] and the end-to-end delay of voice packets [9], respectively. The QoS requirement is given in terms of the mean end-to-end voice packet delay in [9]. However, in [5,6], a stochastic delay requirement is considered in the capacity analysis. The voice sources are treated as constant-rate traffic sources in [5, 6, 11], and on-off traffic sources in [7, 8, 10]. In [9], three different CAC strategies used for voice over IP (VoIP) traffic in wireless networks are mathematically analyzed with respect to centralized CRNs. The authors study the impact of the primary user information on the CAC algorithms in [10]. Different CAC algorithms based on the number of SUs in the system and the total number of users (both the PUs and SUs), respectively, are compared. The call blocking probability and the packet loss probability are considered as the QoS parameters. Developing channel access schemes for voice service support over distributed fully-connected CRNs and the capacities of the developed schemes are presented in [5, 6]. In almost all of the works, the primary network under consideration is time-slotted, and the channel availability for the secondary users can be either independent or dependent in adjacent time-slots, respectively. Most existing studies assume perfect channel sensing by the secondary users, except in [8] which extends the capacity analysis to incorporate the effect of imperfect channel sensing. Different form the previous studies, the voice service support over TV bands is studied in [11]. The minimum number of channels required to be sensed by each SU to support a given traffic load is analyzed using an optimization technique. The channel sensing errors are also incorporated into the optimization problem. Further, deployment of long term evolution (LTE)-Advanced in the industrial, scientific, and medical (ISM) band to resolve some deployment issues is studied in [12]. The authors propose a statistical traffic control scheme to tackle critical challenges of the packet transmission coordination and the radio resources allocation in the network.

The voice capacity studies in the existing works are carried out for single voice packet transmission per time slot per frequency channel. Transmitting multiple voice packets as a composite packet over a time-slot (per channel) improves the network capacity without the requirement of a proportional increment in the channel transmission rate. However, transmitting multiple voice packets as a composite packet can increase the delay jitter, which is not desirable in the context of voice service. There is a trade-off between reducing the packet delay and delay jitter. Furthermore, the existing works are limited to the case that all the voice traffic flows require the same service quality. For example, in [5, 6], all the voice traffic flows have a stochastic delay requirement with the same delay bound and maximum delay bound violation probability. In general, the larger the delay bound, the lower the probability of delay bound violation. Therefore, a larger number of users having a larger delay bound can be accommodated in the system, which gives the possibility to provide the service at a lower cost. However, the larger the delay bound, the lower the service quality. There is a trade-off between the service cost and the service quality. By incorporating different delay requirement parameters (D_{max} and ϵ), the system provides users a choice between service quality and cost, which improves the satisfaction of the users. In the literature, there are only limited works in developing channel access schemes to support voice [6, 7] over CRNs, and much attention has not been payed to the generic channel access schemes in supporting voice traffic.

2.1.1 Motivation and Objectives

Treating the voice source as an on-off traffic source leads to higher system capacities. Further, stochastic delay guarantees are provided in shared wireless networks to provide higher service satisfaction. However, time varying nature of the on-off voice traffic flows and the random nature of the channel availability of the CRNs make the service satisfaction challenging. Therefore, on-off voice capacity has not been studied in the literature providing a stochastic delay guarantee. In this chapter, we study the on-off and constant-rate voice capacity of a centralized CRN with a stochastic delay requirement. A base station may not be available in a CRN due to the cost of installation. Therefore, the node coordination is difficult, and interruptions by the primary users makes it more challenging. In distributed CRNs, efficient channel access schemes are required for the network coordination. The system capacity depends on the efficiency of the channel access schemes, and benchmarks are important to compare the efficiency of the developed schemes. Readily available legacy channel access schemes such as round-robin and slot-ALOHA schemes can be used as the benchmarks, and studying the voice capacity of CRNs with legacy channel schemes is useful for the CR research community. In this chapter, we study the capability of legacy channel access schemes in supporting voice traffic over fully-connected CRNs. In non-fully-connected networks, the larger the number of neighbors associated with target user, the lower the channel time available per user, leading to lower service quality. Therefore, CAC plays an important role in keeping the number of neighbors per user to an acceptable level. In this chapter, we develop CAC algorithms for non-fully-connected slot-ALOHA based CRNs in supporting voice traffic.

2.1.2 Contributions

The contribution of this chapter is five fold: (1) We analyze the on-off and constant-rate voice capacity of a single-channel centralized CRN with FCFS service discipline; (2) We analyze the constant-rate voice capacity of a single-channel distributed fully-connected CRN with slot-ALOHA channel access coordination.

Different from the existing work, we consider the transmission of multiple voice packets in a single time-slot; (3) We analyze the voice capacity of a single-channel distributed fully-connected CRN for round-robin channel access coordination. As the capacity analysis approach used for the slot-ALOHA scheme cannot be used for the round-robin scheme, a new approach is introduced. Further, possible extensions to the analytical models to incorporate sensing errors are discussed; (4) We develop a CAC procedure for a distributed non-fully-connected CRN with slot-ALOHA network coordination, assuming homogeneous single-hop voice traffic flows. The capacity analysis results of the fully-connected network is used to limit the number of calls entering the system; (5) We develop two CAC algorithms for a distributed non-fully-connected slot-ALOHA CRN when the voice traffic flows have different delay requirements. For all the above studies, both dependent and independent channel occupancies of PUs in neighboring time-slots are considered, and the end-to-end delay of voice packets is considered as the QoS parameter. Note that (1) is presented in [13] and (2)–(5) are presented in [14, 15].

2.2   System Model

The system architecture, channel model, voice traffic model, and channel access schemes under consideration are described in this section.

2.2.1   System Architecture

In this study, we consider a centralized CRN with a base station and a distributed fully-connected CRN as illustrated in Fig. 2.1a, b, respectively. The secondary network operates over a time-slotted single-channel primary network, and all the SUs see the same spectrum opportunities[1] (spectrum homogeneous). The secondary network is an overlay CRN in which the SUs access the channel (transmit or receive) only when the PUs are not present. Each SU is equipped with a single transceiver to sense the channel and transmit information packets.

2.2.2   Channel Availability Model

The channel time is partitioned into slots of constant duration T_S. The channel state of each time-slot is either idle (i.e., no primary activities) or busy (i.e., with primary activities). In a time-slot, the state is defined as 0 if the channel is busy,

---

[1]The primary user activities are consistent throughout the network. Therefore, the coverage area of the secondary network should be smaller than the one hop coverage area of the primary network.

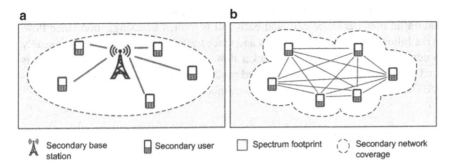

a

b

Secondary base station

Secondary user

Spectrum footprint

Secondary network coverage

Fig. 2.1 (**a**) The centralized CRN with a BS. (**b**) The distributed fully-connected CRN

and is 1 otherwise. The state transition of the channel among adjacent time-slots can be given using a Markov chain as illustrated in Fig. 2.2, where $S_{i,j}$ denotes the transition probability from state i to state j $(i, j \in \{0, 1\})$. This is a widely used method to model the behavior of primary users [5–8, 16] due to its simplicity. The channel state can be independent or dependent among adjacent time-slots. In the independent case, $S_{0,1}=S_{1,1}=p_1$ and $S_{1,0}=S_{0,0}=p_0=1-p_1$. For both the independent and dependent cases $S_{0,1}+S_{0,0}=1$ and $S_{1,0}+S_{1,1}=1$. The channel state is identified by an SU by spectrum sensing. Different sensing techniques are listed in [17] and references there in. For the simplicity of analysis, we assume that the final sensing decisions of the SUs and the BS are free of errors.[2] A time-slot is mainly divided into sensing and transmission phases, and in addition, distributed networks need a contention phase before the transmission phases. An SU transmits only when the channel is at State 1 (available for SUs), and the sensing and transmissions are free of errors.

Fig. 2.2 The channel state transition diagram

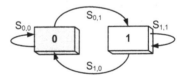

2.2.3 *Voice Traffic Model*

All the SUs are voice nodes, and each voice call is associated with an SU and the BS or two SUs. Each of the two SUs (or an SU and the BS) initiates an independent

---

[2]In reality, sensing errors are inevitable. Under the future extensions of this work, we discuss the effects of the sensing errors on the secondary network and the primary network, respectively.

voice traffic flow to the other. For the simplicity of our analysis, we consider only one traffic flow per voice call, and each call is limited to a single-hop voice flow. In the following, the terms voice call and voice traffic flow are used interchangeably to denote a one-way single-hop packet flow of a voice call, and the term node is used to denote an SU. Widely accepted voice traffic models, namely, the constant-rate voice model [18, 19] and on-off voice model [20, 21] are considered.

2.2.3.1 Constant-Rate Voice Model

A voice node generates a constant-rate traffic flow with a packet inter-arrival time of T_I (normalized to the time-slot duration T_S), and the output of a voice codec is illustrated in Fig. 2.3.

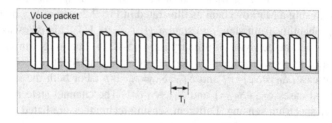

Fig. 2.3 The codec output of a constant-rate voice source

2.2.3.2 On-Off Voice Model

The on-off model is a common voice traffic model used in the VoIP applications [7–10]. The state transition diagram of an on-off voice node is illustrated in Fig. 2.4a, where $Q_{i,j}$ denotes the transition rate from state i to state j ($i, j \in \{0, 1\}$).

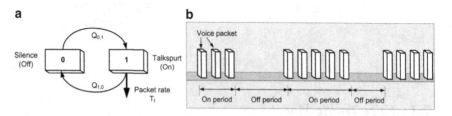

Fig. 2.4 (**a**) The state transition, (**b**) the codec output of an on-off voice source [20]

When a voice node is at the off state, it does not generate any packet; when at the on state (talk spurt), it generates voice packets at a rate of $1/T_I$. An output of a voice codec is illustrated in Fig. 2.4b. When an on-off voice node is at the off state with an empty buffer, the voice node is at its inactive state, and when it is in the on state or off state with a non-empty buffer, it is in its active state.

2.2.3.3 Voice Buffer Management

Each voice source buffers the voice packets until it gets an spectrum (transmission) opportunity. The service requirement for the voice traffic flows in the secondary network is characterized by the end-to-end delay of a voice packet (i.e., from the time that a packet is generated at the source node to the time that it is received at the receiver node). As the packet propagation delay is negligible when compared with the time that a packet spend at the source buffer, the service requirement is given in terms of the queuing delay D (normalized to T_S), from the time that a packet is generated at the source node to the time that it is transmitted from the source node. The stochastic delay requirement is given by [3, 6, 22]

$$P(D > D_{\max}) \le \epsilon \tag{2.1}$$

where $D_{\max}$ (normalized to T_S) and ϵ are the delay bound and maximum delay bound violation probability allowed, respectively, in order to provide satisfactory voice quality. If the delay bound of a voice packet is violated, the packet is dropped without being transmitted. Without loss of generality, we assume integer values for T_I, D, and $D_{\max}$.

2.2.3.4 Voice Capacity

The voice capacity is defined as the maximum number of simultaneous voice calls that can be supported by the system, without violating the delay requirement given in (2.1) for all the admitted calls. In the analysis, it is important to note that the number of voice calls refers to the number of one-way voice calls.

2.2.3.5 Service Disciplines for Voice Traffic

For the centralized network, we consider an ideal scenario with the BS having the queue head waiting times of all the nodes.

- **FCFS service discipline**: The BS schedules packet transmission of the nodes in the available time slots based on the maximum queue head waiting time first principle.

For the distributed (fully-connected) CRN, two legacy channel access schemes, namely, the slot-ALOHA scheme and the round-robin scheme are considered, and the random allocation scheme is used as a benchmark for the performance comparison.

- **Round-robin scheme**: Each node will wait for its channel access right. When a particular node receives the channel access right, it transmits if it has packets in the buffer, or forwards the opportunity to the next node otherwise. Due to the cyclic nature of getting the channel access right, each node accesses the channel in a fair manner. As an approach of realizing the round-robin channel access coordination, a token based scheme [23] or a mini-slot based scheme [6] can be used. There are no packet collisions in the round-robin scheme as a node transmits only when it has the channel access right.
- **Slot-ALOHA scheme**: All the nodes with a non-empty buffer will transmit with a probability ϱ during an idle time-slot. If a collision occurs, each node will re-transmit at the next available time-slot with the same probability.
- **Random allocation**: One node act as a controller and assigns the channel access right to the other nodes randomly.

As the first step, we analyze the capacity of a centralized CRN with ideal information availability, and we extend the study to a distributed CRN with less information for the transmission decision. For simplicity of the analysis, we will only consider networks with stationary nodes.

2.3   Voice Capacity of the Centralized CRN

We analyze the voice capacity of a centralized CRN with a BS to schedule the channel access of each user, as illustrated in Fig. 2.1a. Only single-hop voice communication occurs between voice nodes, and all the voice flows have the same delay requirement. The arrival process and the service process of the centralized system are illustrated in Fig. 2.5, where the outputs of all the voice codecs are either on-off or constant-rate.

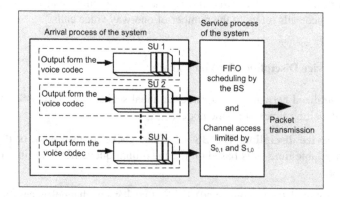

Fig. 2.5 The arrival process and the service process of the centralized system

2.3.1 Service Process Analysis

Since the channel is time-slotted as discussed in Sect. 2.2.2, its service process is a discrete-time process. The service process, $\mu_S(n)$, is defined as the number of packets that can be transmitted in the time-slot n, and is given by $\mu_S(n)=X_S(n)\cdot n_{pk}$, where $X_S(n)$ ($\in \{0,1\}$) is the channel availability index of the time-slot, and n_{pk} is the maximum number of packets that can be transmitted in a time-slot. Service process analysis with respect to the QoS requirement can be carried out using the theory of effective capacity (EC), as discussed in [24–29]. The EC provides the constant arrival rate that can be supported by the system (service process), without violating the required service quality. The EC analysis of a block fading channel [28, 30] can be adopted to analyze our channel by modeling the channel as a single block fading channel with two fading amplitudes (0 and 1). Therefore, the EC of the secondary network with dependent channel availability in adjacent time-slots is given by

$$\zeta_c(\theta) = -\frac{1}{\theta} \ln \left[\frac{S_{0,0} + S_{1,1}e^{-\theta n_{pk}}}{2} + \sqrt{\left(\frac{S_{0,0} - S_{1,1}e^{-\theta n_{pk}}}{2} \right)^2 + S_{0,1}S_{1,0}e^{-\theta n_{pk}}} \right]$$

(2.2)

where, θ depends on the QoS requirement, and it is shown in [25] that $P(D \geq D_{\max}) \approx e^{-\theta D_{\max}}$. In order to satisfy the condition (2.1), the parameter θ should satisfy the condition $\theta \geq \frac{1}{D_{\max}} \ln \left(\frac{1}{\epsilon} \right)$. In order to support a constant arrival rate r with the given delay requirement (2.1), $\delta^* = r\zeta_c^{-1}(r)$ should be satisfy the condition

$$\delta^* \geq \frac{1}{D_{\max}} \ln \left(\frac{1}{\epsilon} \right).$$

(2.3)

The EC of the secondary network with independent channel availability scenario can be obtained by setting $S_{0,1}=S_{1,1}=p_1$ and $S_{1,0}=S_{0,0}=p_0=1-p_1$.

2.3.2 Arrival Process Analysis

The capacity requirement of the constant-rate voice traffic sources remains $1/T_I$ packets/time-slot throughout the duration of the call. However, the capacity requirement of an on-off voice source varies with time. An on-off traffic source can be characterized by the mean, m, variance, ϑ, auto covariance time coefficient, ς, and peak-to-mean ratio, υ, of the traffic flow [31]. The four parameters are given by $m=Q_{0,1}/T_I(Q_{0,1} + Q_{1,0})$, $\vartheta=m(\frac{1}{T_I} - m)$, $\varsigma=1/(Q_{0,1} + Q_{1,0})$, and $\upsilon=(Q_{0,1} + Q_{1,0})/Q_{0,1}$. The four parameters corresponding to the aggregate traffic from N independent sources are given by $m_a=N\cdot m$, $\vartheta_a = N\cdot\vartheta$, $\varsigma_a = \varsigma$, and $\upsilon_a = \upsilon$, respectively. The aggregate traffic flow can be characterized by a two-state Markov modulated Poisson process (MMPP). The MMPP can be characterized by four parameters R_1, R_2, ϕ_1, and ϕ_2, where R_i is the mean rate of the Poisson process in state i, and ϕ_i is the transition rate from state i ($i \in \{1, 2\}$). The four parameters are given by, $R_1 = m_a + \sqrt{\upsilon_a \vartheta_a}$, $R_2 = m_a - \sqrt{\vartheta_a / \upsilon_a}$, $\phi_1 = \upsilon_a / \varsigma_a (1 + \upsilon_a)$, and $\phi_2 = 1/\varsigma_a(1 + \upsilon_a)$ [31]. The capacity requirement of a time varying arrival process considering its service requirements is carried out using the theory of effective bandwidth (EB), as explained in [32,33]. The EB of an arrival process is the required constant service rate in order to satisfy the service quality requirement of the arrival process. The effective bandwidth of the two-state MMPP is given by [31]

$$\zeta_b(\theta) = \frac{\varXi(\varPhi + (e^\theta - 1)\acute{R})}{\theta} \tag{2.4}$$

where $\varPhi$ is the transition rate matrix of the two-state Markov chain of the aggregated traffic flow, $\acute{R} = \mathrm{diag}(R_1, R_2)$, and $\varXi(\cdot)$ gives the largest real eigen value. In order to satisfy the given delay requirement (2.1) using a constant service rate u, $\delta^* = u\zeta_b^{-1}(u)$ should satisfy the condition (2.3) [24]. When both the arrival and service processes are time varying, in order to satisfy the delay requirement (2.1), $\delta^* = \theta^* \zeta_c(\theta^*)$ should be satisfy the condition (2.3), where θ^* is the solution to the equation $\zeta_b(\theta) = \zeta_c(\theta)$ [25]. Note that, in the case of N constant-rate voice traffic sources, $\zeta_b(\delta) = N/T_I$. In order to determine the maximum number of voice sessions that can be supported by the system while satisfying the stochastic delay requirement (2.1), we have to find maximum N which satisfies (2.3).

2.4 Voice Capacity of the Distributed CRN

The constant-rate voice traffic capacity of a distributed fully-connected CRN will be studied in this section under round-robin and slot-ALOHA channel access schemes, and it will be compared with the random allocation scheme. In all three cases, even though the voice buffer of each node acts in the FCFS manner, the system with all the voice nodes as a whole does not behave in a FCFS manner. Therefore, the theory of EB and EC cannot be directly applied to the system, but, to each node. However, due to the complexity of analyzing the service process (of each node), we resort to packet level analysis of the voice buffer of each node.

2.4.1 Slot-ALOHA Scheme

With the initiation of a voice traffic flow, the first packet enters the source buffer becomes the queue-head, and the rest of the packets are buffered behind the queue-head. Whenever the queue-head is successfully transmitted, the next packet with the highest waiting time becomes the new queue-head, χ_{new}. While awaiting for transmission, the waiting time of the queue-head increases with time. However, when a successful transmission occurs, the waiting time of χ_{new} is always lower than that of the queue-head, χ_{old}, which is just being transmitted. The waiting time, D_{new}, of the new queue-head (normalized to T_S), is given by

$$D_{new} = D_{old} - n_s{\cdot}T_I + 1 \tag{2.5}$$

where D_{old} (normalized to T_S) is the waiting time of χ_{old}. The term $n_s{\cdot}T_I$ is due to the n_s inter-arrival times between the arrivals of χ_{old} and χ_{new}, and the constant 1 accounts for the time-slot taken for the transmission of χ_{old}. As the voice packets whose waiting time exceeds the delay bound are dropped, the waiting time of a queue-head stays between 0 and D_{max}. When a packet (queue-head) is dropped due to violation of the delay bound (i.e., $D > D_{max}$), the waiting time of χ_{new} is given by $D_{new}=(D_{max}+1)-T_I$. The queue-head is dropped at the beginning of the time-slot when $D_{old} = D_{max}+1$. The term T_I is due to the inter-arrival time between the χ_{old} and the χ_{new}. In each idle time-slot, a target node with a non-empty buffer transmits with probability ϱ, and a successful transmission occurs if all the other non-target nodes in the network do not transmit. The probability of successful transmission (same as the probability of successful channel access), $P_{S,1}$, in an available time-slot is given by

$$P_{S,1} = \varrho\,(1 - \rho{\cdot}\varrho)^{N-1} \tag{2.6}$$

where ρ is the probability of a node having a non-empty buffer. The product $\varrho{\cdot}\rho$ is the probability of a node transmitting in an idle time-slot. Note that, the probability $P_{S,1}$ does not depend on D. The value of D at the next time-slot depends on the value of D, the state of the channel, and the success or failure of the transmission in the current time-slot. Furthermore, the state of the channel in the next time-slot either does not depend on that of the current time-slot for the independent channel availability scenario, or only depends on the state of the channel in the current time-slot for the two-state channel in Fig. 2.2. Therefore, we can establish a discrete-time Markov chain (DTMC) in which the state (i, j) represents the waiting time of the queue-head and the channel state, respectively, as shown in Fig. 2.6. Since there is no queue-head when the buffer is empty, the negative value of the time remaining until the next packet arrival is considered as the queue-head waiting time. Therefore, D varies from $-(T_I - 1)$ to D_{max}. Theoretical aspects of this approach is discussed in [34]. Furthermore, the DTMC model is similar to the approach given in [6], in analyzing the constant-rate voice capacity of two different cognitive radio MAC protocols. Different from [6], here we consider the transmission of possible multiple (up to n_{pk}) voice packets by a node in a time-slot. The state transition probabilities of the Markov chain are given by

$$P_{(k,i),(k+1,j)} = S_{i,j}, \qquad\qquad\qquad k \in \{-T_I + 1, \ldots, -1\}$$

$$P_{(k,i),(k+1,j)} = (1 - P_{S,i}){\cdot}S_{i,j}, \qquad\qquad k \in \{0, \ldots, D_{max} - 1\}$$

$$P_{(k,i),(k-T_I+1,j)} = (1 - P_{S,i}) \cdot S_{i,j}, \qquad k = D_{\max}$$

$$P_{(k,i),((k \bmod T_I)-T_I+1,j)} = P_{S,i} \cdot S_{i,j}, \qquad k \in \{0, \ldots, (n_{pk} \cdot T_I - 1)\}$$

$$P_{(k,i),(k-n_{pk} \cdot T_I+1,j)} = P_{S,i} \cdot S_{i,j}, \qquad k \in \{n_{pk} \cdot T_I, \ldots, D_{\max}\}$$

where $P_{(k,i),(l,j)}$ denotes the transition probability from state (k,i) to state (l,j) and $i, j \in \{0, 1\}$. Since the channel is not available for the SUs when it is at state 0, $P_{S,0}=0$. As the packets whose waiting time is larger than the delay bound are dropped, the delay bound violation probability, P_e, is equal to the packet dropping probability, given by

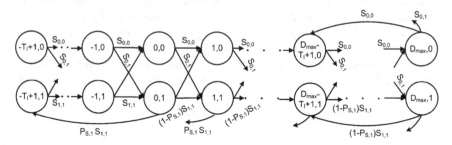

Fig. 2.6 The Markov chain for the queue-head waiting time and channel state pair

$$P_e = \frac{\sum_{j=0}^{1} (1 - P_{S,j}) \cdot \pi_{D_{\max},j}}{P_{S,1} \cdot \sum_{i=0}^{D_{\max}} n_a(i) \cdot \pi_{i,1} + \sum_{j=0}^{1} (1 - P_{S,j}) \cdot \pi_{D_{\max},j}} \qquad (2.7)$$

where $\pi_{i,j}$ is the steady state probability of state (i, j) and $n_a(i)$ is the number of packets that can be transmitted when the queue-head waiting time is i, given by

$$n_a(i) = \begin{cases} \lfloor \frac{i}{T_I} \rfloor + 1, & \lfloor \frac{i}{T_I} \rfloor + 1 < n_{pk} \\ n_{pk}, & \text{otherwise.} \end{cases}$$

The summation $\sum_{j=0}^{1} (1 - P_{S,j}) \cdot \pi_{D_{\max},j}$ represents the mean number of dropped packets and $P_{S,1} \cdot \sum_{i=0}^{D_{\max}} n_a(i) \cdot \pi_{i,1}$ represents the mean number of transmitted packets at the steady state, in a time slot. The capacity analysis problem can be represented as to maximize N with the constraint $P_e \leq \epsilon$. However, the relationship between the probability P_e and N is not straightforward. Therefore, we resort to numerical analysis in calculating the capacity.

We can find the probability $P_{S,1}$ for a given ρ and N by (2.6). Using $P_{S,1}$, the steady state probabilities of the Markov chain can be computed, and thereby the probability of buffer occupancy ρ is given by $\rho = \sum_{j=0}^{1} \sum_{i=0}^{D_{\max}} \pi_{i,j}$. Since probabilities $\pi_{i,j}$ ($i \in \{0, 1, \ldots, D_{\max}\}$ and $j \in \{0, 1\}$) can be represented in terms of ρ, the right hand side (RHS) of the equation also contains ρ. Denote the ρ in RHS as ρ_R and that in the left hand side (LHS) as ρ_L. The value of ρ_L can be computed for different values of ρ_R, and the solution for the equation is the one when $\rho_L = \rho_R$. Then, the probability of delay bound violation P_e can be obtained for a given N. Therefore, the maximum N which satisfies $P_e \le \epsilon$ can be evaluated. The capacity analysis for the independent channel occupancy scenario can be carried out using the preceding method by substituting appropriate values for $S_{i,j}$ ($i, j \in \{0, 1\}$).

2.4.2   Random-Assignment Scheme

As the assignment is random, the probability of successful transmission in an available time-slot is given by $P_{S,1} = 1/N$, and is independent in adjacent available time-slots. Therefore, the same approach used with the slot-ALOHA scheme can be used to analyze the probability of delay bound violation and the voice capacity.

2.4.3   Round-Robin Scheme

The round-robin scheme guarantees that each node gets a packet transmission opportunity in an orderly manner. Whenever the node under consideration (target node) transmits, its next packet transmission does not occur before each non-target node with a non-empty buffer gets an opportunity to transmit. From (2.5), it can be seen that the queue-head waiting time of the target node drops just after a successful transmission. The probability of the target node getting the next transmission opportunity depends on the number of non-target nodes in the network having packets to transmit, the channel availability, and the time elapsed from its previous transmission. Therefore, with the round-robin scheme, the probability of a node getting a packet transmission opportunity is not the same for all D values, and the analysis for the probability of getting a transmission opportunity at the particular D value is not straightforward. Therefore, the Markov chain approach used for the slot-ALOHA scheme cannot be applied for the capacity analysis of the round-robin scheme.

Assuming that the packets of a target node are not dropped until it gets a channel access right (i.e., the packets with the waiting time larger than D_{max} will be dropped at the time the target node gets the channel access right), the range of D is $[0, \infty)$. When the target node gets a channel access right, it will drop $n_d(D)$ and transmit $n_a(D)$ voice packets, where

$$n_d(D) = \begin{cases} 0, & D \le D_{max} \\ \lfloor \frac{D - D_{max} - 1}{T_l} \rfloor + 1, & \text{otherwise} \end{cases}$$

and

$$n_a(D) = \begin{cases} \lfloor \frac{D-n_d(D)T_I}{T_I} \rfloor + 1, & \lfloor \frac{D-n_d(D)T_I}{T_I} \rfloor + 1 \leq n_{pk} \\ n_{pk}, & \text{otherwise.} \end{cases}$$

After transmitting the $n_a(D)$ packets, the D of the queue-head decreases by $(T_I \cdot n_a(D) - 1)$ time-slots. Then, it increases by a random number of time-slots until the next channel access. With N voice calls in the system, for a target node, the waiting time of the queue-head at the time of packet transmission depends on the waiting time of the queue-head at the previous packet transmission and the number of time-slots required to provide a transmission opportunity to each of the $N - 1$ non-target nodes. If the number of time-slots in the shortest possible round-robin cycle is larger than or equal to the number of time-slots between two successive packet arrivals, the target source buffer will always be non-empty when it receives a transmission opportunity. As the shortest possible round-robin cycle is equal to the number of nodes in the network, N, the condition to have a non-empty buffer when a source node receives a transmission opportunity can be expressed as $N \geq T_I$. Therefore, the randomness will only be due to the channel availability, not due to the number of nodes with a non-empty buffer.

As the waiting time D at the next packet transmission depends only on that of the current packet transmission, but not on the previous packet transmissions, a DTMC can be developed with the state representing the queue-head waiting time at the time of packet transmission. With the waiting time D in $[0, \infty)$, the state space of the DTMC lies in the same range, making it an infinite-state DTMC. The Markov chain is illustrated in Fig. 2.7, where $P_{i,j}$ is the transition probability from state i to state j ($i, j \in \{0, 1, 2, \ldots\}$).

For a single-channel CRN with $N \geq T_I$, the state transition probabilities, $P_{i,j}$, of a target node is given by $P_{i,j} = P\left(\sum_{z=0}^{N-1} X_z = r\right)$, if $r \geq N$, and 0, otherwise, where Z is the number of nodes to access the channel before the target node gets the channel access right, X_Z is the number of time-slots required to reduce the

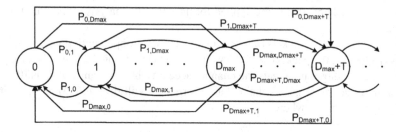

Fig. 2.7 The DTMC for the queue-head delay at the time of packet transmission with round-robin channel access

node number from Z to $Z - 1$,[3] and $r = j - (i - (n_d(i) + n_a(i)) T_I)$ is the elapsed number of time-slots between adjacent channel access opportunities. The number of time-slots X_Z ($Z \in \{0, 1, .. N - 1\}$) are independent and identically distributed. When the channel availability for SUs in adjacent time-slots is independent, the state (the number Z) transition for a node is illustrated in Fig. 2.8. When there are N source nodes in the system and they all have packets to transmit, it is impossible for a target node to have its next transmission opportunity within $N - 1$ adjacent time-slots from its current transmission. Therefore, $P_{i,j} = 0$ for $r < N$. In order to have $r - 1$ time-slots ($r \geq N$) between two successive transmission opportunities, the target node should transmit at the rth time-slot, and the rest of the $N - 1$ non-target nodes should transmit during the first $r - 1$ time-slots. In other words, exactly N out of the r time-slots should be idle and, out of the N idle time-slots, $N - 1$ should be in the first $r - 1$ time-slots. Therefore, the probability $P_{i,j}$ is given by the negative binomial distribution. The state transition probability, $P_{i,j}$, for an independent channel occupancy scenario of PUs is given by

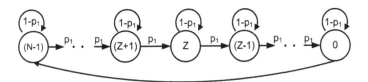

Fig. 2.8 State transition of Z for independent channel occupancy of PUs

$$P_{i,j} = \begin{cases} \binom{r-1}{r-N} p_1^N (1 - p_1)^{r-N}, & \text{if } r \geq N \\ 0, & \text{otherwise.} \end{cases} \tag{2.8}$$

When the channel availability for SUs are dependent among adjacent time-slots, the state Z is divided into two states named Z_1 and Z_2, where a node enters state Z through state Z_1 (initial state), and enters state Z_2 if the channel is not available when it is in state Z_1. The state transition diagram of a node is illustrated in Fig. 2.9. As explained earlier, $P_{i,j} = 0$ for $r < N$. If there are exactly N time-slots in between successive transmissions of the target node, all N time-slots should be available for the SUs.[4] Having $r > N$ time-slots between successive transmissions means that the channel has been idle for N time-slots and busy for $r - N$ time-slots. The state transition probabilities, $P_{i,j}$, for a dependent channel occupancy scenario of PUs is given by

---

[3] A non-target node with channel access right requires X_Z time-slots to obtain a channel opportunity and transmit its packets.

[4] Being in state 1 (idle state), the channel should remain in state 1 for N successive time-slots.

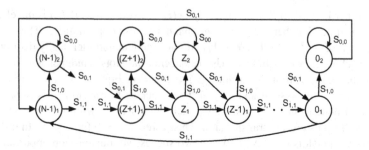

Fig. 2.9 State transition of Z for dependent channel occupancy of PUs

$$
P_{i,j} = \begin{cases} \sum_{l=1}^{\min(r-N,N)} \binom{N}{N-l} S_{1,1}^{N-l} S_{1,0}^{l} \cdot \binom{r-N-1}{l-1} S_{0,1}^{l} S_{0,0}^{r-N-l}, & \text{if } r > N \\ S_{1,1}^{N}, & \text{if } r = N \\ 0, & \text{otherwise.} \end{cases}
\tag{2.9}
$$

In (2.9), when $r > N$, there must be at least one transition from state 1 to state 0. The term $\binom{N}{N-l} S_{1,1}^{N-l} S_{1,0}^{l}$ represents the probability of having l state 1 to state 0 transitions out of all the transitions occur in the N idle time-slots. In order to have N idle time-slots, l state 0 to state 1 transitions are required in the remaining $r - N$ time-slots. The term $\binom{r-N-1}{l-1} S_{0,1}^{l} S_{0,0}^{r-N-l}$ represents the probability of having l state 0 to state 1 transitions in exactly $r - N$ time-slots. Since the DTMC has a countably infinite number of states, it is truncated to $D_{\max} + k \cdot T_I$ states for simplicity of analysis, where k (≥ 1) is a small integer. The delay bound violation probability, P_e, is approximately given by

$$
P_e \simeq \frac{\sum_{i=0}^{D_{\max}+k \cdot T_I} n_d(i) \cdot \pi_i}{\sum_{i=0}^{D_{\max}+k \cdot T_I} (n_d(i) + n_a(i)) \pi_i}
\tag{2.10}
$$

where π_i is the steady state probability of state i. The terms $\sum_{i=0}^{D_{\max}+k \cdot T_I} n_d(i) \cdot \pi_i$ and $\sum_{i=0}^{D_{\max}+k \cdot T_I} n_a(i) \cdot \pi_i$ represent the mean number of dropped packets and transmitted packets, respectively, at the steady state in a time-slot. The system capacity $N_{\max}$ is the maximum N which satisfies the relation $P_e \leq \epsilon$. The larger the N, the larger the P_e. The minimum P_e, P_e^*, that can be analyzed by (2.10) is for the minimum N, N^*. As $N \geq T_I$, $N^* = T_I$. Thus, the capacity can be evaluated for an ϵ value larger than P_e^*.

Capacity analysis of a fully-connected network is the first step of developing a call admission control algorithm. As we evaluate the maximum number $N_{\max}$ of simultaneous voice traffic flows that can be supported by the system without violating the delay requirement, the call admission control can be carried out by limiting the number of traffic flows in the network to $N_{\max}$.

2.5 Call Admission Control

When the slot-ALOHA scheme is used for the channel access control, collisions occur due to simultaneous transmissions of a target source node and the neighboring source nodes associated with the target receiver node. The larger the number of neighboring source nodes associated with a target receiver, the higher the chances of collisions, which leads to a lower successful transmission probability, $P_{S,1}$, of the target source node (or traffic flow). The lower the probability $P_{S,1}$, the longer the waiting time of packets in the buffer and the probability P_e of delay bound violation. Therefore, in order to keep the probability P_e within a desired limit, the number of calls admitted to the system should be controlled.

2.5.1 CAC for Homogeneous Voice Traffic

In Sect. 2.4.1, we analyze the maximum number, $N_{\max}$, of homogeneous voice traffic flows that can be carried out by a slot-ALOHA fully-connected network. Therefore, $N_{\max}$ is the maximum number of homogeneous voice source nodes that can be associated with a target receiver node. In a non-fully-connected network, each receiver node is associated with a number of source nodes. The packet transmission of a new source node increases the collisions at its associated receiver nodes, leading to a reduction in the successful transmission probability of the said receiver nodes. Therefore, to satisfy the delay requirement of the ongoing and incoming traffic flows, it is required to control the admission of new calls based on the number of source nodes associated with each receiver node (including that of the incoming call). A CAC procedure, $P1$, based on the number of neighboring nodes can be explained as follows. Denote the source and receiver nodes of the new call by target source (ω_s) and receiver (ω_r) nodes, respectively, and the set of neighboring receiver nodes of ω_s and source nodes of ω_r by $\in G_{\omega_s}$ and G_{ω_r}, respectively. Let N_{i_r} be the number of neighboring source nodes of receiver node i_r ($\in \{G_{\omega_s} \cup \omega_r\}$). It is required to limit N_{i_r} of each receiver node i_r ($\in \{G_{\omega_s} \cup \omega_r\}$) to a maximum of $N_{\max}$. Therefore, ω_s should listen to its neighbors i_r ($\in G_{\omega_s}$) and get the information N_{i_r}. At the same time, ω_r should listen to its neighbors and find N_{ω_r}. If the condition $N_{i_r} \leq N_{\max}$ can be satisfied for all i_r ($\in \{G_{\omega_s} \cup \omega_r\}$), the new call is admitted to the system, and rejected otherwise. As $N_{\max}$ is a function of ϱ, the non-fully-connected network must use the same (ϱ, $N_{\max}$) pair which used with the fully-connected network.

The capacity of a fully-connected network is under the assumption of homogeneous voice traffic. However, the capacity analysis of the fully-connected network is no longer valid for non-homogeneous voice traffic. The validity of the $N_{\max}$ used in this procedure no longer holds, and a new approach is required for the CAC of non-homogeneous voice traffic over non-fully-connected CRNs.

2.5.2   CAC for Non-homogeneous Voice Traffic

Majority of the existing CAC strategies developed for non-cognitive ad hoc networks consider only the first order statistics such as average waiting time, and are based on standard queuing analysis by using the Little's theorem. Further, there are some existing works on CAC in non-cognitive networks based on stochastic QoS guarantees using the theory of effective bandwidth and its dual effective capacity [3, 22, 35]. All of these works are for homogeneous/non-homogeneous traffic flows with the same delay requirement. Based on this idea, we can develop a CAC algorithm for non-fully-connected CRNs as a bench mark. However, analysis of the effective capacity of the service process of an SU is not straightforward as it depends on the channel access scheme. The approach used in Sect. 2.3.1 to analyze the effective capacity of the CRN can be adopted to analyze that of the service process of each node.

The packet buffer of each source node act in the FCFS service discipline. Therefore, in order to satisfy the delay requirement of voice packets, the effective capacity of the service process of each source node should be larger than the constant arrival rate. A successful packet transmission from a target source node occurs whenever there are no collisions at the target receiver node. Therefore, the service process of the target source node is governed by the transmissions of the neighboring source nodes of the target receiver node. The effective capacities of the discrete-time service process for independent and dependent channel availability scenarios can be obtained by (2.2). In a particular time-slot (**irrespective of its availability**), define the state of the service process of a target source node as follows: If a successful transmission occurs during the time slot, the source node is in state 1, and state 0 otherwise. The effective capacities for independent and dependent channel availability cases are given by

$$\zeta_c(\theta) = -\frac{1}{\theta} \ln \left[\frac{F_{0,0} + F_{1,1} e^{-\theta \cdot n_{pk}}}{2} + \sqrt{\left(\frac{F_{0,0} - F_{1,1} e^{-\theta \cdot n_{pk}}}{2} \right)^2 + F_{0,1} F_{1,0} e^{-\theta \cdot n_{pk}}} \right].$$

where, $F_{i,j}$ $(i, j \in \{0, 1\})$ is the transition probability of a node from state i to state j. In an **available time-slot** define the state of transmission (transmission state) as follows; If a transmission is successful, the source node is in state 1, and state 0 otherwise. Consider a Markov chain in which the state is represented by the channel state and transmission state pair which consist of three states $(1,1)$, $(1,0)$, and $(0,0)$. Denote the state transition probability matrix of the Markov chain by $\tilde{F}$. The node is at state 1 if both the transmission state and the channel state are 1, and state 0 otherwise. The state transition probability matrix F of the service process a source node can be obtained using the state transition probability matrix $\tilde{F}$. The condition to satisfy the delay requirement of a voice call i is given by $\delta^* \zeta_c(\delta^*) \geq \frac{1}{D_{\max}} \log \left(\frac{1}{\epsilon} \right)$, where δ^* is the solution to the equation $\zeta_c(\delta) = \frac{1}{T_I}$.

This condition can be given in the form $\zeta_c\left(\delta_{min}^*\right) \geq \frac{1}{T_I}$, where $\delta_{min}^*=\frac{T_I}{D_{max}}\log\left(\frac{1}{\epsilon}\right)$. In the distributed non-fully-connected network scenario, the probability $P_{S,1}$ of a target source node ω_s is given by

$$P_{S,1} = \Pi_{i_s \in G_{\omega_r}} \varrho_{\omega}(1 - \rho_i \cdot \varrho_i) \tag{2.11}$$

where ϱ_j and ρ_j are the transmission probability given that the buffer is non-empty and the probability of having a non-empty buffer of source j_s ($\in \{G_{\omega_r} \cup \omega_s\}$), and ω_r is the target receiver node. However, the evaluation of ρ_i is not straightforward as it depends on the transmissions of the neighboring source nodes of receiver i_r. Therefore, rather than evaluating the exact value of ρ_i, we investigate the possibility of obtaining a close upper bound for the value of ρ_i. From the DTMC illustrated in Fig. 2.6, it can be seen that the delay bound violation probability P_e of a constant-rate voice traffic flow and the probability ρ of a voice buffer being non-empty, monotonically decrease with the successful transmission probability $P_{S,1}$. Therefore, the delay requirement $P_e \leq \epsilon$ can be transformed to $P_{S,1} \geq P_S^*$ or $\rho \leq \rho^*$, where P_S^* is the $P_{S,1}$ value at $P_e = \epsilon$ and ρ^* is the ρ value at $P_{S,1} = P_S^*$. The variation of ρ and P_e with $P_{S,1}$, and the relationship of P_S^*, ρ^*, and ϵ are illustrated in Fig. 2.10. As long as the existing source nodes satisfy the delay

Fig. 2.10 Variation of ρ and P_e with $P_{S,1}$ for $D_{max}=100$, $T_I=10$, $n_{pk}=5$, $P_{01}=0.8$, $P_{10}=0.2$, and $\epsilon=0.01$

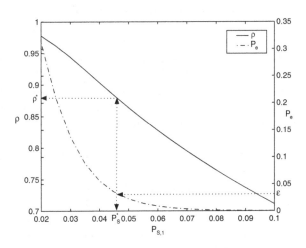

requirement $P_e \leq \epsilon$, the probability ρ is upper bounded by ρ^*. Therefore, instead of using $P_{S,1}$, we substitute $P_{S,1}^*=\Pi_{i_s \in G_{\omega_r}} \varrho_{\omega}(1 - \rho_i^* \cdot \varrho_i)$ ($\leq P_{S,1}$) in (2.11). When the system supports non-homogeneous voice traffic flows with different delay bounds, let C denote the set of all voice traffic classes in the network. Each voice traffic class c ($\in C$) has unique delay bound $D_{max}(c)$, $P_S^*(c)$, and $\rho^*(c)$ values. Therefore, ρ_i^* and ϱ_i of $P_{S,1}^*$ should be replaced by their respective values of the traffic class c_i as $\rho^*(c_i)$ and $\varrho(c_i)$, where $\varrho(c_i)$ is the default ϱ value for the traffic class c_i. Denote the source and receiver nodes of the incoming call, ω, as the target source (ω_s) and receiver (ω_r) nodes. In order to make sure that the delay requirements of all ongoing calls and the new call are satisfied, effective capacities of each receiver node i_r ($\in \{G_{\omega_s} \cup \omega_r\}$) should be larger than the packet arrival rate $1/T_I$. The benchmark CAC algorithm based on the effective capacity is given in algorithm A1.

Data : $C_i = \{c_j : j \in G_{i_r}\}$
Result: $\hat{C}_i$

1   $\hat{C}_i \leftarrow \emptyset$;
2   **repeat**
3   randomly select c_k from C;
4      $C \leftarrow C - \{c_k\}$;
5   **if** $i == \omega$ **then**
6         $P_{S,1} \leftarrow \varrho(c_k) \Pi_{j \in G_{\omega_r}} (1 - \rho^*(c_j) \cdot \varrho(c_j))$;
7   **else**
8         $P_{S,1} \leftarrow \varrho(c_i) \Pi_{j \in G_{i_r}} (1 - \rho^*(c_j) \cdot \varrho(c_j))(1 - \rho^*(c_k) \cdot \varrho(c_k))$;
9   **end**
10     $\delta^*_{min} \leftarrow \frac{T_I}{D_{max}(c_i)} \log\left(\frac{1}{\epsilon}\right)$;
11   **if** $\zeta(\delta^*_{min}) \geq \frac{1}{T_I}$ **then**
12        $\hat{C}_i \leftarrow \{\hat{C}_i \cup c_k\}$;
13   **end**
14 **until** $C == \emptyset$;
15 Exit;

Algorithm 1: CAC algorithm based on the effective capacity

Each receiver node in the network should run the algorithm and identify the set $\hat{C}_j$ ($j \in \{G_{\omega_s} \cup \omega_r\}$). The set $\hat{C}_\omega$ and $\hat{C}_i$ ($i \in G_{\omega_s}$) are the set of voice classes that can be admitted by ω_r, and to the neighborhood of an existing receiver node i_r, respectively, without violating the delay requirement of the existing and incoming voice calls. The new source and receiver nodes listen to the channel and identify the set of voice classes $C_\omega = \bigcap_{i_r \in \{G_{\omega_s} \cup \omega_r\}} \hat{C}_i$ that can admit call ω. If $C_\omega = \emptyset$, call ω cannot be admitted to the system. The effective bandwidth/capacity approach can be applied to different types of traffic by evaluating the effective bandwidth [24, 25] of the source traffic and the effective capacity of the service process via modeling the source buffer occupancy at the packet level. However, this approach is computationally complex due to the requirement of calculating the effective capacity at run-time. It is possible to introduce a less complex approach for the CAC for non-homogeneous voice traffic using the relationship of $P_{S,1}$, ρ, and P_e.

Based on Fig. 2.10, guaranteeing $P_{S,1} \geq P_S^*(c)$ guarantees $\rho \leq \rho^*(c)$. Therefore, if the probability $P_{S,1}$ of source i_s ($\in G_{\omega_r}$) satisfies $P_{S,1} \geq P_S^*(c_i)$, the inequality $\varrho_\omega \Pi_{i_s \in G_{\omega_r}} (1 - \rho_i \cdot \varrho_i) \geq \varrho_\omega \Pi_{i_s \in G_{\omega_r}} (1 - \rho^*(c_i) \cdot \varrho_i)$ always stands. Provided that $P_{S,1} \geq P_S^*(c_i)$ for all $i_s \in G_{\omega_r}$, the delay requirement of the incoming call can be guaranteed by choosing a proper ϱ_ω value for its source ω_s, which satisfies $\varrho_\omega \Pi_{i_s \in G_{\omega_r}} (1 - \rho^*(c_i) \cdot \varrho_i) \geq P_S^*(c_\omega)$. However, as discussed in Sect. 2.5.1, the admission of a new source node increases the probability P_e of delay bound violation of each source i_s, where i_s is the corresponding source node of i_r ($\in G_{\omega_s}$).

Therefore, it is required to guarantee that $P_{S,1}$ values of the said source nodes and the new source node are kept above their respective $P_S^*(c_j)$ ($j_r \in \{G_{\omega_s} \cup \omega_r\}$) values by making sure that the following conditions are met respectively

$$\varrho_j \cdot \prod_{i_s \in G_{j_r}} (1 - \rho^*(c_i) \cdot \varrho_i) \geq P_S^*(c_j), \quad \forall \, j_r \in G_{\omega_s},$$

and

$$\varrho_\omega \cdot \prod_{i_s \in G_{\omega_r}} (1 - \rho^*(c_i) \cdot \varrho_i) \geq P_S^*(c_\omega) \qquad (2.12)$$

where the LHSs of (2.12) are always less than or equal to $P_{S,1}$. The expressions of the LHSs of (2.12) can be evaluated using γ_j ($= \varrho_j \cdot \prod_{i \in G_{j_r}} (1 - \rho^*(c_i) \cdot \varrho_i)$) and c_j obtained from the neighboring receiver nodes of the new source node, and γ_ω ($= \prod_{i \in G_{\omega_r}} (1 - \rho^*(c_i) \cdot \varrho_i)$) obtained from the new receiver node. The CAC algorithm based on the relationship among P_e, $P_{S,1}$, and ρ is given in Algorithm 2.

In the algorithm, parameter ϱ_{min} is the minimal ϱ value which satisfies the first inequality in (2.12), ϱ_{max} is the maximal ϱ value which satisfies the second inequality in (2.12), and $\acute{\beta}$ is the transmission probability selection parameter. Algorithm 2 searches for ϱ_{min} and ϱ_{max} by increasing ϱ_ω from 0 to 1 in a step size ϱ_s. The smaller the ϱ_s, the higher the accuracy of $\varrho_{min}\varrho_{max}$ values. However, the smaller the ϱ_s, the larger the number of iterations required to get the results, leading to a larger processing time. If the algorithm outcome is to admit the call, it needs to choose a ϱ_ω value ($\varrho_{min} \leq \varrho_\omega \leq \varrho_{max}$) for the transmissions of the new source node. The probability $P_{S,1}$ of the new source node and corresponding source nodes of its neighboring receiver nodes will vary depending on the chosen ϱ_ω value. Therefore, a particular $\acute{\beta}$ value should be selected for the network to obtain a ϱ_ω ($= \varrho_{min} + \acute{\beta}(\varrho_{max} - \varrho_{min})$) value, such that the network capacity is maximized. This can be carried out by trial and error method off-line.

2.6 Numerical Results

Computer simulations are carried out to evaluate the accuracy of the capacity analysis of the given channel access schemes and to investigate the performance of the two CAC algorithms. In order to depict the primary user activities, the channel is made on and off according to the dependent and independent channel occupancy statistics of PUs. The voice traffic classes used in the analysis are given in Table 2.1 Note that all the time durations are normalized to T_S. The typical values of the on and off durations of an on-off voice source are around 320 ms and 640 ms, respectively [20, 36]. However in [36], it is shown that these durations are dependent on the factors such as conversation topics and situations of voice calls. The probability of delay bound violation, P_e, is obtained by the ratio of the

Data : $\Gamma=\{(\gamma_i, c_i) : i \in \{G_{\omega_s} \cup \omega_r\}\}$
Result: ϱ_ω, Admit the call or block the call

1 $\kappa \leftarrow 0$;
2 $\varrho_\omega \leftarrow \varrho_s$;
3 **while** $\varrho_\omega \leq 1$ **do**
4   **if** $\gamma_\omega \cdot \varrho_\omega \geq P_S^*(c_\omega)$ **then**
5          $\varrho_{min} \leftarrow \varrho_\omega$;
6   Go to 10;
7   **end**
8      $\varrho_\omega \leftarrow \varrho_\omega + \varrho_s$;
9 **end**
10 **while** $\varrho_\omega \leq 1$ **do**
11     $H_g \leftarrow G_{\omega_s}$;
12   **repeat**
13   randomly select j_r from H_g;
14         $H_g \leftarrow H_g - \{j_r\}$;
15   **if** $\gamma_j \left(1 - \rho^*(c_\omega) \cdot \varrho_\omega\right) < P_S^*(c_j)$ **then** Go to 21;
16   **until** $H_g == \emptyset$;
17     $\kappa \leftarrow 1$;
18     $\varrho_{max} \leftarrow \varrho_\omega$;
19     $\varrho_\omega \leftarrow \varrho_\omega + \varrho_s$;
20 **end**
21 **if** $\kappa == 1$ **then**
22     $\varrho_\omega \leftarrow \varrho_{min} + \acute{\beta}(\varrho_{max} - \varrho_{min})$;
23   Admit the call;
24 **else**
25   Block the call;
26 **end**
27 Exit;

Algorithm 2: CAC algorithm based on the successful transmission probability

Table 2.1 The voice traffic classes used in the simulations

Voice traffic class	Notation	Traffic type	Mean on duration	Mean off duration	T_I	D_{max}
Class 0	c_0	On-off	320	640	10	30
Class 1	c_1	Constant-rate	–	–	10	30
Class 2	c_2	Constant-rate	–	–	10	100
Class 3	c_3	Constant-rate	–	–	10	250

number of dropped packets (at a source node due to the violation of delay bound) to the total number of packets generated by the source node. Each simulation result is obtained by averaging the results of 10 simulation runs each having a duration of 10^6 time-slots.

2.6.1 Capacity Analysis

Consider homogeneous voice traffic flows of class c_i ($i \in \{0, 1, 2, 3\}$). While keeping N constant during a simulation run, the probability P_e is obtained for a particular channel access scheme and channel statistics. Starting from $N = 2$, we increase N by one for each simulation run and the resultant probability P_e is compared with ϵ to obtain $N_{\max}$, which satisfies $P_e \leq \epsilon$.

2.6.1.1 Centralized Network

Figure 2.11a, b show the variation of $N_{\max}$ with n_{pk} in a centralized network with FCFS service discipline and different channel availability statistics for on-off (class c_0) and constant-rate (class c_1) voice traffic, respectively. The results demon-

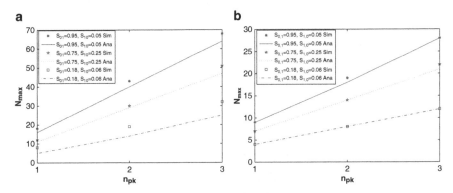

Fig. 2.11 Variation of $N_{\max}$ with n_{pk} for on-off voice traffic in a (**a**) centralized, (**b**) distributed network with FCFS service discipline

strate that the analytical results match closely with the simulation results. However, the analytical results stay slightly below the simulation results due to the conservative nature of the theory of effective bandwidth and its dual, effective capacity. The capacity of the system increases with the number of voice packets that can be transmitted in a single time-slot (per channel), n_{pk}. However, a proportional increment in the channel rate is not required to increase the n_{pk}. Therefore, capability of transmitting multiple packets in a single-time-slot can have a considerable impact on the system capacity. In the on-off voice sources under consideration, the duration of talkspurts are only one third of the call duration. The number of voice packets generated in an on-off voice traffic flow is approximately one third of that of a constant-rate voice traffic flow. Therefore, with the given FCFS service discipline (the ideal scenario), the on-off voice traffic provides more than twice the capacity of the constant-rate voice traffic.

2.6.1.2   Distributed Network

Consider homogeneous voice traffic flows of class c_2. Figure 2.12a–c show the variation of P_e with N obtained from numerical analysis and simulations with slot-ALOHA, random allocation, and round-robin channel access schemes, respectively, for different channel availability statistics with $n_{pk} = 4$. The results demonstrate that the P_e obtained form simulation match well with the analytical results in all three scenarios. The system capacity N_{max} is given by the maximum N having P_e less than ϵ (0.01 in our simulation). Further, it can be observed that the higher the mean channel availability, $p_1 = S_{0,1}/(S_{0,1} + S_{1,0})$, the higher the system capacity in all three cases.

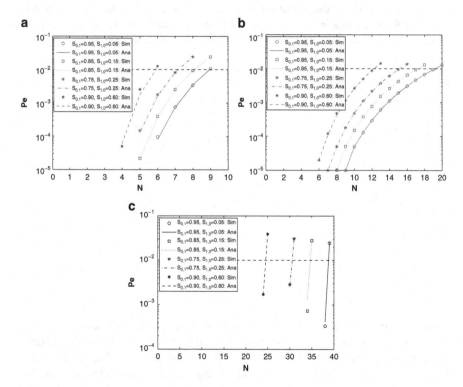

Fig. 2.12 Variation of P_e with N for a fully-connected network with (**a**) slot-ALOHA, (**b**) random allocation, (**c**) round-robin channel access

Figure 2.13 shows the variation of N_{max} with n_{pk} for all three channel access schemes having p_1 constant in 0.95.

It is observed that the higher the mean channel availability and n_{pk}, the higher the capacity in all three channel access schemes. Further, the round-robin scheme provides the highest system capacity and slot-ALOHA provides the worst system capacity, when compared with the other two. Note that the overhead required

Fig. 2.13 Variation of N_{max} with n_{pk} for a fully-connected network for all three channel access schemes

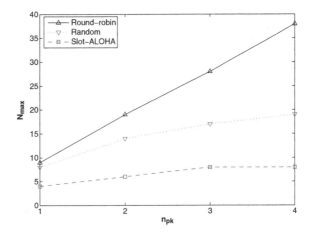

for the establishment of the round-robin scheme is much higher than that of the slot-ALOHA scheme, as explained in Sect. 2.2.3, which is neglected in the simulation. Furthermore, it is observed that the rate of increment of system capacity with n_{pk} in the round-robin scheme is higher than the random allocation and slot-ALOHA schemes. In the slot-ALOHA and random allocation schemes, the probability of transmission is irrespective of the buffer occupancy of packets. However, in the round-robin scheme, there is a higher probability to transmit when the waiting time of the queue-head is larger (i.e., when there are more number of packets in the buffer), which allows a node to transmit a larger number of voice packets during a transmission than in the other two schemes. Therefore, the mean number of packets transmitted during a channel access opportunity is smaller in the slot-ALOHA and random allocation schemes than that in the round-robin scheme, which explains that the latter has a higher rate of capacity improvement with n_{pk}. The system capacity with the round-robin scheme is similar to that of centralized network with FCFS service discipline. Since, the round-robin scheme does not need the packet timing information, it can be considered as a promising candidate for voice service support over CRNs.

Figure 2.14 shows the variation of N_{max} with T_{off} for all three channel access schemes having p_1 constant in 0.8 and $n_{pk} = 4$.

The results demonstrate that the longer the T_{off}, the lower the system capacity even though the mean channel availability remains constant. The longer the T_{off}, the longer the duration of the busy periods of the channel from the viewpoint of the SUs, leading to longer the durations of packet waiting times. This increases the probability of delay bound violation of the voice packets. Even if the channel available duration is longer (corresponding to longer T_{on}) it is not possible to transmit the packets which have already violated the delay bound. Therefore, the longer the T_{off}, the lower the system capacity. This shows the importance of considering the state transition probabilities of the channel in analyzing the system capacity rather than considering the mean channel availability.

Fig. 2.14 Variation of
$N_{\max}$ with T_{off} for a
fully-connected network for
all three channel access
schemes

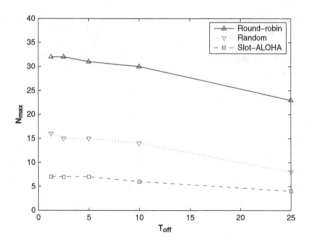

2.6.1.3 Discussion

In our work, we assume ideal channel sensing (i.e., error free detection of primary activities). However, sensing errors are inevitable in practical scenarios. There are two types of sensing errors, namely, missed detection (an SU or the BS does not detect the presence of a PU) and false alarm (an SU or the BS detects presence of a PU while the PUs is not present) [37]. The missed detections lead SUs to transmit simultaneously with the PUs, causing interference. In order to establish the CRN, the probability, P_{md}, of missed detection has to be controlled below a certain threshold to minimize the interference with the PUs. Despite the interference, the SU transmission can be successful. The false alarms reduce the channel utilization of the SUs. Therefore, the probability, P_{fa}, of false alarm has to be minimized to improve channel utilization of the SUs. From the viewpoint of an SU (or the BS), the channel availability differs from the true state of the channel due to the presence of the sensing errors. Denote the channel state from the viewpoint of an SU (or the BS) as the virtual channel state. The virtual channel state transition probabilities can be obtained using the channel state transition probability matrix and the error probabilities P_{md} and P_{fa}, respectively. We can incorporate the effect of the sensing errors into our capacity analysis after some efforts on modification of the Markov chains.

When the sensing errors are present, packet transmissions of the SUs in the centralized network depend on the virtual channel state of the BS. In order to incorporate the sensing errors into the capacity analysis, the channel state transition probabilities in (2.2) should be replaced by the virtual channel state transition probabilities. In the fully-connected slot-ALOHA network, the packet transmission of an SU depends on the channel sensing errors. Therefore, the state transition probabilities of the DTMC in Fig. 2.6 should be modified to incorporate the sensing errors. The modifications to the state transitions are illustrated in Fig. 2.15, where $P_{S,md}$ is the probability of successful transmission given the occurrence of a

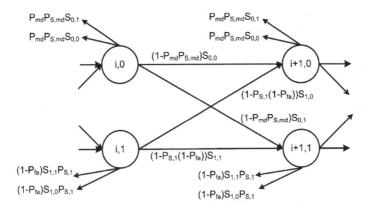

Fig. 2.15 The modifications to the state transitions of the DTMC in Fig. 2.6 to incorporate sensing errors

missed detection and $P_{S,1} = \varrho(1 - \rho\varrho(1 - P_{fa}))^{N-1}$ is the successful transmission probability given the channel is available. The transition probabilities $P_{md} P_{S,md} S_{0,j}$ and $(1 - P_{fa}) P_{S,1} S_{1,j}$ are due to successful transmissions given the channel is not available (transmission being successful due to missed detection) and available (when there is no false alarm), respectively, where $j \in \{0, 1\}$. The transition probability $(1 - (1 - P_{fa}) P_{S,1}) S_{1,j}$ and $(1 - P_{md} P_{S,md}) S_{0,j}$ are due to non occurrence of packet transmissions when the channel is available and not available, respectively, where $j \in \{0, 1\}$. Similar to that given in Sect. 2.4.1, the delay bound violation probability can be evaluated by finding the steady state probabilities of the DTMC. The sensing errors can be incorporated into the capacity analysis of the round-robin scheme by modifying the state transition probabilities of the DTMC in Fig. 2.9, and the modification depends on the mechanism used to establish the round-robin scheme. With the above, it is clear that the sensing errors can be incorporated in our capacity analysis with slight modifications in the state transition probabilities of the DTMCs.

In a multiple channel network, an SU will either select a channel and sense for availability or sense all the channels and select an available channel. Two approaches lead to different successful transmission probabilities. We can extend our capacity analysis approach for a multiple channel CRN by evaluating the corresponding successful transmission probabilities. The service quality is given in terms of queuing delay D. The QoS requirement can be relaxed by defining the service quality in terms of the mean of queuing delay, $E[D]$.

As mobile video is predicted to generate most of the mobile traffic growth through 2005 [1], it is important to study video streaming over the CRNs. As given in [38], video frames are generated in burst according to a coding and compression algorithm, and each video burst consists of a number of video packets (with a pdf given by negative binomial distribution). The video clips are grouped into a small number of shot classes depending on the burst size, and a video traffic flow is modeled my a Markov modulated Gamma process in [39]. The author also analyze the EB of the video traffic flow for a maximum data loss rate of 10^{-2}. Therefore, the number of video traffic flows that can be supported by the centralized FCFS system can be studied using the EC evaluated in Sect. 2.3.1 and the EB approach in [39]. For the distributed networks, the possibility to carry out a packet level analysis of the source buffer can be studied, given the statistics of the video bursts and the probability of channel access. Further research is necessary to model the source buffer state using a Markov chain, and to analyze the packet dropping probability using the steady state probabilities as given in Sect. 2.4.1.

2.6.2   Call Admission Control

For the performance comparison of the CAC procedure (P1) and two CAC algorithms (A1 and A2), we consider a CRN with homogeneous voice traffic. For the performance comparison of algorithms A1 and A2, we consider a network with both traffic classes, where new call arrivals are equally likely to be of class c_2 or c_3. The network coverage area of each voice source/receiver node is a circle with a radius of unit length. The inter-arrival time of voice calls is exponentially distributed, and the location of source nodes is uniformly distributed in a square network area. Ten different data sets are generated, each containing 8,000 samples of source and receiver location and call inter arrival time. In order to compare the two algorithms, 10 different simulation runs were carried out for each algorithm using the generated data sets over a constant network area. As the network is non-fully-connected, the system capacity depends on the coverage area of the network. We saturate the network with voice calls to obtain the maximum number of voice calls that can be supported by the system, and obtained the results for different network coverage areas.

Figure 2.16 shows the comparison of the network capacity (with the 95% confidence interval) of class c_2 voice calls using procedure P1 and algorithms A1 and A2. The CAC procedure P1 outperforms the algorithms A1 and A2 when the mean channel availability is lower, and the algorithm A2 outperforms the other two when the mean channel availability is higher. The algorithm A2 opportunistically chooses the probability ϱ at the instance of call admission whereas P1 has a fixed ϱ value. Therefore the opportunistic ϱ selection may choose different ϱ values for different calls leading to a probability $P_{S,1}$ which is just enough to satisfy the admission criterion (ϱ can vary from ϱ_{min} to ϱ_{max}). The lower the channel availability, the larger the P_S^*. The larger the P_S^*, the lower the tolerance for admitting a new call and vice versa. Therefore, lower channel availability can lead to lower capacities when the ϱ selection is opportunistic as in A2. The performance of algorithm A1 always stays below A2 due to the conservative nature of the theory of effective capacity. The required information from the neighboring nodes and the calculation complexity of P1 is less than those of A1 and A2. Therefore, the

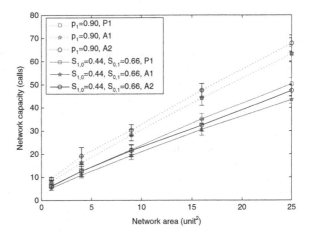

Fig. 2.16 Variation of the network capacity with the network area for procedure $P1$ and algorithms A1 and A2

procedure P1 can be a better choice over A1 and A2 at low channel availability, and A2 can be a better choice over P1 and A1 for a network with homogeneous voice traffic at high channel availability.

Figure 2.17 shows the variation of the average network capacity with the network area, using algorithms A1 and A2 for the two equally likely voice traffic classes.

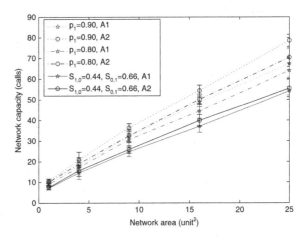

Fig. 2.17 Variation of the network capacity with the network area for algorithms A1 and A2 for a network with two voice traffic classes

The results demonstrate that algorithm A2 is a better choice over A1. The average network capacity with the mixture of two voice traffic classes is higher than that for voice class c_2. The relaxed QoS requirement of voice traffic class c_3 allows more calls to be admitted. Clearly, there is a trade-off between the number of calls in the systems and the service quality, as expected. Algorithm A2 can be extended to other contention based channel access schemes (e.g. IEEE 802.11 RTS/CTS based) and traffic types, given that P_e and ρ monotonically decrease with P_s, and the channel contention is independent over adjacent time-slots.

2.7   Summary

In this chapter, we have studied the voice capacity and call admission control for CRNs. A stochastic delay requirement, both independent and dependent channel availability statistics, and different channel access schemes are considered. The on-off voice capacity of a centralized CRN is studied for the FCFS service discipline using the theories of effective bandwidth and its dual effective capacity. The analytical results appear to be slightly lower than the simulation results due to the conservative nature of the theory of effective bandwidth. Further, it is observed that the silent-suppressed (on-off) voice sources provides more than twice the capacity over that of constant-rate voice traffic. The existing DTMC model is modified to analyze the capacity of slot-ALOHA scheme, and a new DTMC model is developed to analyze the capacity of round-robin scheme in supporting constant-rate voice traffic over distributed fully-connects CRNs. It is shown that the round-robin scheme performs better than the other two schemes, and the capacity is very close to that of the FCFS scheme used in the centralized network. Therefore, the round-robin scheme is a better choice in fully-connected networks to support voice traffic, and it can be established using a token based scheme or a mini-slot based scheme as explained in [6, 23]. Further, the maximum number of voice packets that can be transmitted in a time-slot and the mean channel unavailable duration have a significant impact on the system capacity. The longer the mean channel unavailable durations the lower the system capacity, even when the mean channel availability remains unchanged. In order to relax the assumption of perfect channel sensing, possible extensions to the analytical models to incorporate the sensing errors are discussed.

We use the capacity analysis results of the fully-connected network to limit the number of neighboring users of each target voice user in non-fully-connected CRNs with slot-ALOHA network coordination. It is only applicable for constant-rate voice sources with the same delay requirement. However, having long delay bounds and large delay bound violation probabilities can increase the system capacity. Therefore, we develop two new CAC algorithms to support voice sources with different delay requirements (different delay bound and maximum delay bound violation probability). It is shown that the loner the delay bound, the larger the system capacity. In other words, the lower the required service quality, the higher the system capacity. A low quality service can be priced at a lower cost than a high quality service. Giving the users an option to choose the required service quality can increase the level of user satisfaction.

References

1. Cisco visual networking index: global mobile data traffic forecast update, 2010–2015. Tech. rep., Cisco (2011)
2. Allot mobile trends: global mobile broadband traffic report. Tech. rep., Allot (2010)

3. Abdrabou, A., Zhuang, W.: Stochastic delay guarantees and statistical call admission control for ieee 802.11 single-hop ad hoc networks. IEEE Trans. Wireless Commun. **7**(10), 3972–3981 (2008)
4. Cai, L., Xiao, Y., Shen, X., Cai, L., Mark, J.: Voip over wlan: voice capacity, admission control, qos, and mac. Int. J. Commun. Syst. **19**(4), 491–508 (2006)
5. Wang, P., Niyato, D., Jiang, H.: Voice service support over cognitive radio networks. In: Proc. IEEE ICC (2009)
6. Wang, P., Niyato, D., Jiang, H.: Voice service capacity analysis for cognitive radio networks. IEEE Trans. Wireless Commun. **59**(4), 1779–1790 (2010)
7. Lee, H., Cho, D.H.: Voip capacity analysis in cognitive radio system. IEEE Commun. Lett. **13**(6), 393–395 (2009)
8. Lee, H., Cho, D.H.: Capacity improvement and analysis of voip service in a cognitive radio system. IEEE Trans. Veh. Technol. **59**(4), 1646–1651 (2010)
9. Cruz-Perez, F.A., Rivero-Angeles, M.E., Hernandez-Valdez, G., Castellanos-Lopez, S.L.: Joint call and packet level performance analysis of cac strategies for voip traffic in wireless. In: Proc. IEEE GLOBECOM (2011)
10. Castellanos-Lopez, S.L., Cruz-Perez, F.A., Rivero-Angeles, M.E., Hernandez-Valdez, G.: Impact of the primary resource occupancy information on the performance of cognitive radio networks with voip traffic. In: Proc. IEEE CROWNCOM (2012)
11. Koufos, K., Ruttik, K., Jantti, R.: Voice service in cognitive networks over the tv spectrum. IET Commun. Mag. **6**(8), 991–1003 (2012)
12. Lien, S., Chen, K.C.: Statistical traffic control for cognitive radio empowered lte-advanced with network mimo. In: Proc. IEEE INFOCOM, pp. 80–84 (2011)
13. Gunawardena, S., Zhuang, W.: Voice capacity of cognitive radio networks. In: Proc. IEEE ICC (2010)
14. Gunawardena, S., Zhuang, W.: Voice capacity of cognitive radio networks for both centralized and distributed channel access control. In: Proc. IEEE GLOBECOM (2010)
15. Gunawardena, S., Zhuang, W.: Capacity analysis and call admission control in distributed cognitive radio networks. IEEE Trans. Wireless Commun. **10**(9), 3110–3120 (2012)
16. Zhao, Q., Tong, L., Swami, A., Chen, Y.: Decentralized cognitive mac for opportunistic spectrum access in ad hoc networks: A pomdp framework. IEEE J. Select. Areas Commun. **25**(3), 589–600 (2007)
17. Bagwari, A., Singh, B.: Comparative performance evaluation of spectrum sensing techniques for cognitive radio networks. In: Proc. IEEE CICN, pp. 98–105 (2012)
18. Wang, L., Chen, A., Wei, D.: A cognitive mac protocol for qos provisioning in overlaying ad hoc networks. In: Proc. IEEE CCNC, pp. 1139–1143 (2007)
19. Timmers, M., Dejonghe, A., der Perre, L.V., Catthoor, F.: A distributed multi-channel mac protocol for cognitive radio networks with primary user recognition. In: Proc. IEEE CROWNCOM, pp. 216–223 (2007)
20. Schwartz, M.: Broadband Integrated Networks. Prentice Hall (1996)
21. Chang, C.S.: Performance guarantees in communication networks. Springer London (2000)
22. Abdrabou, A., Zhuang, W.: Statistical qos routing for ieee 802.11 multi-hop ad hoc networks. IEEE Trans. Wireless Commun. **8**(3), 1542–1552 (2009)
23. Wang, P., Zhuang, W.: A token based scheduling scheme for wlans supporting voice/data traffic and its performance analysis. IEEE Trans. Wireless Commun. **7**(5), 1708–1718 (2008)
24. Wu, D., Negi, R.: Effective capacity: A wireless link model for support of quality of service. IEEE Trans. Wireless Commun. **2**(4), 630–643 (2003)
25. Wu, D., Negi, R.: A distributed multi-channel mac protocol for cognitive radio networks with primary user recognition. In: Proc. IEEE BROADNETS, pp. 527–536 (2004)
26. Shakkottai, S.: Effective capacity and qos for wireless scheduling. IEEE Trans. Automatic Control **53**(3), 749–761 (2008)
27. Wu, D., Negi, R.: Effective capacity channel model for frequency-selective fading channels. J. Wireless Networks **13**(3), 299–310 (2006)

28. Tang, J., Zhang, X.: Quality-of-service driven power and rate adaptation over wireless links. IEEE Trans. Wireless Commun. **6**(8), 3058–3068 (2007)
29. Liu, L., Chamberland, J.: On the effective capacities of multiple antenna gaussian channels. In: Proc. IEEE ISIT, pp. 2583–2587 (2008)
30. Chang, C.S..: Stability, queue length, and delay of deterministic and stochastic queuing networks. IEEE Trans. Automatic Control **39**(5), 913–931 (1994)
31. Fan, Z., Mars, P.: Effective bandwidth approach to connection admission control for multimedia traffic in atm networks. IEEE Elec. Lett. **32**(16), 1438–1439 (1996)
32. Hossain, E., Bhargava, V.: Cognitive wireless communication networks. Springer US (2007)
33. Wu, D., Negi, R.: Effective capacity: a wireless link model for support of quality of service. IEEE Trans. Wireless Commun. **2**(4), 630–643 (2003)
34. Servi, L.D.: D/g/1 queues with vacations. Operations Research **34**(4), 619–629 (1986)
35. Lin, L., Fu, H., Jia, W.: An efficient admission control for ieee 802.11 networks based on throughput analysis of unsaturated traffic. In: Proc. IEEE GLOBECOM, pp. 3017–3021 (2005)
36. Deng, S.: Traffic characteristics of packet voice. In: Proc. IEEE ICC, vol. 3, pp. 1369–1374 (1995)
37. Jiang, H., Lai, L., Fan, R., Poor, H.V.: Optimal selection of channel sensing order in cognitive radio. IEEE Trans. Wireless Commun. **8**(1), 297–307 (2009)
38. Song, W., Zhuang, W.: Performance analysis of probabilistic multipath transmission of video streaming traffic over multi-radio wireless devices. IEEE Trans. Wireless Commun. **11**(4), 1554–1564 (2012)
39. Song, W.: Resource reservation for mobile hotspots in vehicular environments with cellular/wlan interworking. EURASIP J. Wireless Communications and Networking (2012)

Chapter 3
Service Response Time of Elastic Data Traffic

3.1 Introduction

As we have discussed in Chap. 2, various studies on conversational and streaming type traffic flows over CRNs have been carried out over the recent years. So far, little attention is paid to the performance analysis of request-response type services such as web browsing over CRNs. The impact of primary user activities on traffic congestion and the economic interaction between secondary user and primary network operators are studied in [1] and [2], respectively, when the SUs are data users. This type of services does not require strict QoS as in conversational or streaming services, but has a moderate service requirement in the form of response time.

Most of the resource allocation/scheduling works in CRNs mainly focus on throughput optimization/fairness, and they do not deal with any specific data file length distributions or response time as a performance metric. The data service is a non-real-time service, where the rate of flows adjusts to fill available bandwidth [3]. Therefore, the data service is also called elastic data service. A performance analysis of elastic data traffic in non-cognitive networks is carried out in [3]. Different bandwidth sharing techniques based on maximum throughput, min-max fairness, proportional fairness, and weighted fairness are considered in the analysis. The MRT evaluation of elastic data traffic flows is studied for cellular/WLAN integrated networks in [4]. The network supports streaming and elastic data traffic flows, and the data files are served in processor sharing service discipline. A MRT approximation for the SRPT service discipline under a heavy traffic condition is given in [5] (and references there in). In all these works, the short-term mean channel rate available for a data user does not vary with time, and therefore the long-term mean channel rate is used for the response time analysis. However, in CRNs the channel availability for SUs varies with time due to the interruptions by the PUs (bursty PU traffic), and the short-term mean channel availability deviates from the long-term mean channel availability. Therefore, the effect of the transmission interruptions caused by the PUs should be considered in the analysis. In [6], the

S. Gunawardena and W. Zhuang, *Modeling and Analysis of Voice and Data in Cognitive Radio Networks*, SpringerBriefs in Computer Science,
DOI 10.1007/978-3-319-04645-7__3, © The Author(s) 2014

mean throughput and delay of transmission control protocol (TCP) and constant bit rate connections are analyzed for CRNs with on-off PU behaviors. However, there aren't many research efforts devoted on the performance analysis of elastic data traffic over CRNs. From the viewpoint of the SUs in a CRN, the available channel time can be considered as a service with break downs. The expected queue lengths and related operating characteristics of a queuing station with breakdown are studied in [7], which can also be applied in the context of CRNs. Further, the relationship between the queuing station with breakdown and a single server queuing system with preemptive priorities is also studied. However, the work in [7] is limited to the FCFS service discipline, which is not always the best service discipline. In all these works, the mean of the response time is considered as the service quality parameter, due to the complexity of analyzing its probability distribution.

3.1.1   Motivation and Objectives

Elastic data traffic falls into the non-real-time traffic category, and the MRT is an important measure to provide a satisfactory service quality. The service interruptions due to the presence of PUs increase the response time of elastic data traffic flows operating over CRNs. The length of requested files vary according to the file length distribution (Weibull distribution is common for Web traffic [8]), which also has a significant impact on the MRT. Different service disciplines can provide different priorities to the service requests based on their file lengths. For example, the SPTWP service discipline provides a higher priority to the requests with short file lengths than the requests with long file lengths, whereas the PS service discipline provides equal opportunity to all the requests. Therefore, different service disciplines result in different MRTs. The longer the response time (i.e., the longer the duration a user waits to start reading/viewing the file), the lesser the user satisfaction. Choosing the correct service discipline based on the service time requirement (or the file length) distribution can reduce the MRT of the files. Therefore, it is important to study the relationship between the MRT and the different system parameters such as the channel availability statistics (PU behavior), data session arrival process, and the service time requirement, respectively, for a given service discipline. The MRT results can be used in developing CAC algorithms for elastic data traffic. The objective of this chapter is to study the MRT of elastic data traffic over CRNs with different service disciplines, and find the impact of different system parameters on the MRT.

3.1.2   Contributions

The contribution of this chapter is three fold: (i) We derive mathematical expressions for the MRT of elastic data traffic service operating over a single channel time-slotted centralized CRN with three service disciplines, namely, shortest processor time without preemption [9], shortest processor time with preemption, and shortest remaining processing time, in comparison with the processor sharing service discipline. The PU activities are considered to have an on-off behavior with on and off durations following geometric distributions; (ii) We compare the MRTs of all four service disciplines under different data traffic load conditions, and demonstrate that the SPTNP is a better choice over the PS service discipline for a heavy traffic load condition; (iii) We compare the MRTs of all four service disciplines under Weibull distributed service time requirements with different tail properties, and demonstrate that the preemption reduces the MRT when the service time requirement (data file size) follows a heavy tailed distribution. The response time analysis can be used for call admission control to ensure service satisfaction. To the best of our knowledge, this is the first study of the service response time for the elastic data traffic under the service disciplines with service interruptions for a CRN. Note that (i)–(iii) are presented in [9, 10].

3.2 System Model

The system architecture, channel model, elastic data traffic model, and channel access schemes under consideration are described in this section.

3.2.1 System Architecture

In this study, we consider a centralized CRN with a base station as illustrated in Fig. 3.1. The secondary network operates over a time-slotted single-channel primary network, and the secondary network is an overlay network which is spectrum homogeneous. The BS and the SUs use a low-rate control channel to transmit control packets. Each SU is equipped with a single transceiver to sense the channel and transmit information packets.

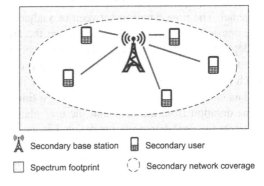

Fig. 3.1 The centralized CRN with a BS

3.2.2   Channel Availability Model

The channel model is similar to the one discussed in Sect. 2.2.2, in which the time is partitioned into slots of constant duration. The channel state in each time-slot is defined as 0 if the channel is busy, and is 1 otherwise. The state transition of the channel among adjacent time-slots can be illustrated using a two state Markov chain as illustrated in Fig. 3.2, where $S_{i,j}$ denotes the transition probability from state i to state j $(i, j \in \{0, 1\})$.

Fig. 3.2 The channel state transition diagram

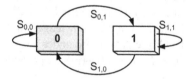

The channel state can be independent or dependent among adjacent time-slots. The mean channel availability and unavailability are given by $p_1 = S_{0,1}/(S_{0,1}+S_{1,0})$ and $p_0 = S_{1,0}/(S_{0,1}+S_{1,0})$, respectively. The channel state is identified by an SU by spectrum sensing and the SUs transmit the sensing decision to the BS for the final channel state decision. A time-slot is mainly divided into sensing and transmission phases. The BS can transmit information packets only when the channel is at State 1 (available for SUs), and the sensing and transmissions are free of errors. The appearance of a PU and the end of an idle period (no PU is active) is denoted as service interruption to the secondary network.

3.2.3   Elastic Data Traffic Model

Web browsing nodes are considered as elastic data traffic sources in the network. The network supports a large number of data users, and the session requests arrive according to a Poisson process with mean λ. Each data session consists of one file request from an SU, and it places the request at the BS via the control channel. The time duration between two adjacent file requests from the same user are considered to be very long such that the arrival time of the second request is independent of that of the first one. After receiving the file request, the BS transfers the file to the particular SU according to a pre-assigned service discipline. During each available time-slot, only one data user is being served and the size, L_{pk}, of a data chunk (packet) transferred during a time-slot is same for all time-sots. The time duration from the time that the user places the file request at the BS until it receives the final data packet is denoted as the response time, T_R. We use the terms service request to denote a file request of an SU. The response time depends on the STR of the request. The STR depends on the length, L, of the requested data file, and L_{pk}. The lengths of requested files are independently and identically distributed with a Weibull distribution which is common for Internet data traffic [8, 11]. The probability density function (PDF), $f_L(\cdot)$, of file length L is given by

$$f_L(x) = \frac{\alpha}{\beta}\left(\frac{x}{\beta}\right)^{\alpha-1} e^{-\left(\frac{x}{\beta}\right)^{\alpha}}, \quad \alpha > 0, \beta > 0, x > 0 \tag{3.1}$$

where α and β are the shape parameter and the scale parameter, respectively. The shape parameter governs the tail heaviness of the Weibull distribution. The smaller the α the longer the tail of the STR distribution. Upon a service request, the whole data file is available at the BS without any delay. The MRT, $E[T_R]$, of a data file is considered as the QoS parameter[1] of the elastic data traffic flows, and the QoS requirement is given by

$$E[T_R] \leq \bar{T}_{R,\max} \tag{3.2}$$

where $\bar{T}_{R,\max}$ is the maximum allowable MRT to provide satisfactory service quality.

3.2.4 Service Disciplines for Data Traffic

When an active SU places a service request at the BS, the BS transmits packets of the requested file based on SPTNP, SPTWP, SRPT, and PS service disciplines. During each available time-slot, only one data user is being served.

- **SPT service discipline without preemption**: When a new service request (target request) arrives at the BS, it is served without any delay if there is no user currently being served (current user), or is placed in a waiting queue otherwise. Once the current user is being served, the request with the shortest STR in the queue will be served. If the channel becomes unavailable (interrupted) during the service of the current user, the service will be halted for the duration of the interruption and resumed after the interruption. If the target request arrives in an interruption period while there is no current request waiting to resume its service, it will be placed in the waiting queue until the channel becomes available, and the user with the lowest STR will be served. This type of interruption is referred to as an idle interruption.
- **SPT service discipline with preemption**: The target request preempts the current user if the **original** STR of the current user is larger than that of the target request. If the target request arrives in an interruption period, the request with the lowest original STR will be served after the interruption.

---

[1] The mean of the response time is considered as the service quality parameter due to the complexity of analyzing its probability distribution.

- **SRPT service discipline**: The target request preempts the current user if the **remaining** STR of the current user is larger than that of the target request. If the target request arrives in an interruption period, the request with the lowest remaining STR will be served after the interruption.
- **PS service discipline**: Users are served in a round-robin manner. If the channel is available in a particular time-slot, the BS transmits a data packet to the user who has the channel access right (current user) and the channel access right is given to the next user in a round-robin order for the next time-slot. However, if the channel is not available in the given time-slot, the current user keeps its channel access right until the next available time-slot. When a new service request arrives, it will be placed last at the round-robin order. In this way, each user gets a fair channel access opportunity, regardless of their service time requirements.

The service disciplines can be directly applied to a fully-connected network with distributed channel access control, where all the SUs are connected to each other by one-hop links. Each time-slot may consist of channel sensing, random contention, and data transmission periods [12]. Further research is necessary to develop an efficient distributed channel access scheme and to apply the service disciplines.

3.3   Response Time Analysis

Denote the queue in which the service requests are placed before they are served for the first time as the waiting queue. In other words a service request is placed in the waiting queue before the transmission of its first data packet. In order to transmit the first data packet of a request, the BS places the service request at the service queue and it remains at the service queue for the rest of its service. The packet transmission and channel available/unavailable durations are in discrete-time due to the time slotted nature of the primary network. However, for analysis tractability, the channel availability/busy durations, and the service time requirements are considered to be in continuous time. That is, the channel availability and busy durations are exponentially distributed with mean $1/\lambda_I$ $(= 1/S_{1,0})$ and $E[I]$ $(= 1/S_{0,1})$, respectively. Without loss of generality, the size of a data packet, L_{pk}, is considered as one unit. Therefore, the PDF of the service time requirement, τ, is $f_\tau(x) = f_L(x)$. The mean STR, $E[\tau]$ is given by $E[\tau] = \int_0^\infty v f_\tau(v) dv$. The cumulative distribution function (CDF) of STR τ is denoted by $F_\tau(x) = \int_0^x f_\tau(v) dv$.

3.3.1   SPT Service Discipline Without Preemption

The conditional MRT, $E[T_R|\tau = t]$, of a target request given its STR, τ, equal to t is [7, 13, 14]

$$E[T_R|\tau = t] = E[W_t] + E[X_t] \qquad (3.3)$$

where T_R is the response time of a target request, W_t is the waiting time of the target request from the time that the user places the service request until the BS starts transmitting the first data packet (i.e., duration that the request spend at the waiting queue), and X_t is the service time of the target request from the time that the BS starts transmitting the first data packet until it transmits the final data packet, which includes the interruption periods during the service (i.e., duration that the request spend at the service queue). As a new request arrives at the system with the STR exactly equal to that of the current user (i.e., $\tau=t$) occurs with a negligible probability, we neglect it for the clarity of presentation.

3.3.1.1 Categorization of Service Requests and Channel Time

Service requests and channel time are categorized based on the service time requirements and type of the request currently using the channel, respectively, as illustrated in Fig. 3.3.

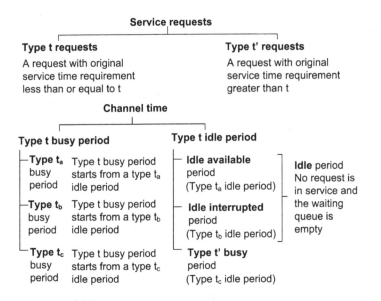

Fig. 3.3 Categorization of service requests and channel time

Define a type t request (type t' request) as a service request with original STR smaller (greater) than t [13]. Type t busy period is defined as a continuous time period during which type t requests are using the channel or being interrupted while using the channel. An illustration of Type t busy periods for a channel without interruptions is given in [13].

If the channel is not in a type t busy period, it is in a type t idle period. A type t busy period starts from a request with STR less than t which arrives during a type t idle period as illustrated in Fig. 3.4, and it lasts until there is no type t request in the system waiting to be served. A type t idle period is divided into two parts, namely, type t′ busy (type t_c idle) period and idle period. A type t′ busy period is a continuous time period during which requests with original STR greater than t are using the channel or being interrupted while being served. An idle period is categorized into idle available (type t_a idle) period and idle interrupted (type t_b idle) period based on the channel availability. An idle available period is a continuous time period during which the channel is available and is not being used by any user. An idle interrupted period is an interruption period which starts from an idle available period.

A type t busy period is categorized into three (type t_a, type t_b, and type t_c) busy periods based on the arrival period of the initiating type t request. The type t_a, type t_b, and type t_c busy periods initiate due to the arrival of a type t request during an idle available, idle interrupted, and type t′ busy period, respectively. A type t_a busy period is initiated at the time of a type t request arrival during an idle available (type t_a idle) period. However, a type t_b busy period initiates just after the completion of an idle interruption (type t_b idle) period, and a type t_c busy period is initiated just after the completion of current type t′ request (type t_c idle period). Examples for the initiation of type t_a, type t_b, and type t_c busy periods are given in Fig. 3.4.

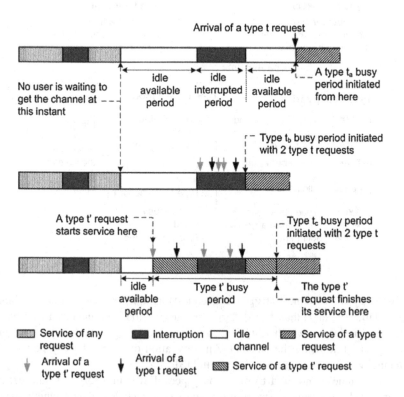

Fig. 3.4 Initiation of type t_a, type t_b, and type t_c busy periods

If a new service request (target request) with STR equal to t arrives during the service time of a request (current request) with original STR smaller than t, the new service request falls into a type t busy period; otherwise, the target request falls into a type t idle period.

3.3.1.2 Target Request Arriving in a type t Idle Period

If a target request arrives in an idle available (type t_a idle) period, it will get the channel access immediately. Therefore, the response time $T_R = X_t$.

If the target request arrives in an idle interrupted (type t_b idle) period, first it waits until the interruption duration finishes. Further, any type t request arrivals during the idle interruption create a type t busy period. If so, the target request needs to wait until the end of the type t busy period to get the service. Therefore, the mean waiting time, $E[W_t] = E[\varphi_{idle,int}] + E[T_{busy,t_b}]P(N_b \geq 1)$, where $\varphi_{idle,int}$ is the residual time of the idle interruption, T_{busy,t_b} is the duration of a type t_b busy period, and N_b is the number of type t requests arrived during an idle interrupted period.

If the target request arrives in a type t' busy (type t_c idle) period, it waits first until the end of current service which is a type t' request. Further, any type t request arrivals during the service time (including the interruptions) of the current user create a type t busy period, and the target request needs to wait until the end of the type t busy period to get the channel access. Therefore, the mean waiting time, $E[W_t] = E[\varphi_{t'}] + E[T_{busy,t_c}]P(N_c \geq 1)$, where $\varphi_{t'}$ is the residual time of a type t' request, T_{busy,t_c} is the duration of a type t_c busy period, and N_c is the number of type t requests arrived during the service time (including the interruption duration) of a type t' request.

3.3.1.3 Target Request Arriving in a type t Busy Period

If the target request arrives in a type t busy period, it waits until the end of the type t busy period to get the service. Therefore, the waiting time $E[W_t] = E[\varphi_{busy,t}]$, where $\varphi_{busy,t}$ is the residual time of the type t busy period. A type t busy period can be any of type t_a, type t_b, and type t_c busy periods. A summary of the waiting times of a target request falling into different time periods are given in Table 3.1. The first type t request that arrives in an idle available period initiates a type t_a busy period. Therefore, the number of type t requests at the initiation of a type t_a busy period is one. The probability of an incoming request initiating a type t_a busy period is $P_{idle,av}F_\tau(t)$, where $P_{idle,av}$ is the probability of the target request arriving in an idle available period.

A type t request with the shortest STR that arrives in an idle interrupted period initiates a type t_b busy period just after the interruption. However, at the initiation of the type t_b busy period, there may be more than one type t request waiting to get service. As the target request has to wait until all the type t

Table 3.1 Waiting time of a target request arriving in different time periods

Time period	Waiting time	SPTNP	SPTWP/ SRPT
Type t_a idle	No waiting time. Immediately receives service (accesses the channel). $E[W_t]=0$.	Yes	Yes
Type t_b idle	Wait until the interruption is over. If there are any type t arrivals during the interruption, a type t_b busy period is generated, wait until the end of the type t_b busy period. $E[W_t]=E[\varphi_{idle,int}]+E[T_{busy,t_b}]P(N_b \geq 1)$.	Yes	Yes
Type t_c idle	Wait until the service completion of the current (type t') user. If there are any type t arrivals during the service of current user, a type t_c busy period is generated, wait until the end of the type t_c busy period. $E[W_t]=E[\varphi_{t'}]+E[T_{busy,t_c}]P(N_c \geq 1)$.	Yes	No
Type t busy	Wait until the end of the ongoing type t busy period. $E[W_t]=E[\varphi_{busy,t}]$.	Yes	Yes

requests finish their service, we can treat any of the type t requests arriving in an idle interrupted period as the initiating request of the type t_b busy period. The probability that an incoming service request initiates a type t_b busy period is then $P_{idle,int}F_\tau(t)$, where $P_{idle,int}$ is the probability of the target request arriving in an idle interrupted period.

Similarly, the probability that a request initiates a type t_c busy period is $P_{busy,t'}F_\tau(t)$, where $P_{busy,t'}$ is the probability of the target request arriving in a type t' busy period. Therefore, the probability, $P_{I,t}$, that an incoming request initiates a type t busy period is given by $P_{I,t}=F_\tau(t)[1 - P_{busy,t}]$, where $P_{busy,t}$ is the probability of the target request arriving in a type t busy period. The conditional MRT, $E[T_R|\tau = t]$ is given by

$$E[T_R|\tau = t] = \left(E[\varphi_{idle,int}] + E[T_{busy,t_b}]P(N_b \geq 1)\right)P_{idle,int} \quad (3.4)$$
$$+ \left(E[\varphi_{t'}] + E[T_{busy,t_c}]P(N_c \geq 1)\right)P_{busy,t'}$$
$$+ E[\varphi_{busy,t}]P_{busy,t} + E[X_t].$$

The MRT of a target request can be evaluated by averaging (3.4) over the PDF, $f_\tau(\cdot)$, of the STR.

3.3.1.4 Mean type t Busy Period

The duration, $T_{busy,t}$, of a type t busy period is given by

$$T_{busy,t} = \sum_{k=0}^{\infty} T_k \tag{3.5}$$

where $T_k = T_k' + I_k'$, $I_k' = \sum_{j=0}^{N_{I,k}} I_j$ $(k \geq 1)$, $T_k' = \sum_{j=0}^{N_{y,k-1}} Y_{k-1,j}$ $(k \geq 1)$, $N_{y,k-1}$ is the number of type t request arrivals during the period T_{k-1}, $Y_{k-1,j}$ is the STR of the jth $(j \in \{0, 1, \ldots, N_{y,k-1}\})$ type t request arrival during T_{k-1}, $N_{I,k}$ is the number of interruptions during the period T_k', and I_j is the duration of the jth $(j \in \{0, 1, \ldots, N_{I,k}\})$ interruption arrived in T_k' $(I_0, Y_{k-1,0} = 0$ by definition). The time duration T_0' in T_0 is the total STR of the $N_{y,0}$ initiating type t requests of the type t busy period. An example for a type t busy period is illustrated in Fig. 3.5. Similar to the analysis given in [15], it can be shown that

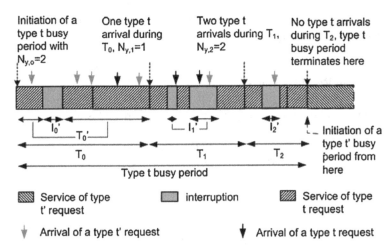

Fig. 3.5 An example of a type t busy period initiated with $N_{y,0} = 2$

$$E[T_{busy,t}] = E\left[\sum_{k=0}^{\infty} T_k\right] = \frac{E[T_0]}{(1 - b_t)} \tag{3.6}$$

where $E[T_0] = E[T_0' + \sum_{j=0}^{N_{I,0}} I_j]$, $b_t = (1 + \lambda_I E[I])\lambda_t E[Y_t]$ is the fraction of time having a type t busy period, λ_t is the mean arrival rate of type t users, and $E[Y_t]$ $(= E[\tau | \tau < t])$ is the expected value of the STR of type t requests. It can be shown that

$$E[T_0] = \frac{1}{p_1} E[N_{y,0}] E[Y_t], \tag{3.7}$$

where $p_1 = 1/(1 + \lambda_I E[I])$ is the mean channel availability. The mean waiting time of a type t request due to type t_b busy period which falls during an idle interrupted period, $E[T_{t_b}]$, can be evaluated similar to (3.6) as

$$E[T_{t_b}] = \frac{b_t}{(1 - b_t)} E[I] \tag{3.8}$$

where $\lambda_t E[I]$ is the mean number of type t request arrivals in an idle interrupted period. The mean duration $E[T_{t_b}]$ can be given by $E[T_{t_b}] = E[T_{busy,t_b}] P(N_b \geq 1) + 0 \cdot P(N_b = 0)$, and therefore, $E[T_{t_b}] = E[T_{busy,t_b}] P(N_b \geq 1)$. Further, the mean waiting time of a type t request due to type t_c busy period which falls during the service time of a type t' request, $E[T_{t_c}]$ ($= E[T_{busy,t_c}] P(N_c \geq 1)$), can be given similar to (3.8) as

$$E[T_{t_c}] = \frac{b_t}{(1 - b_t)} E[X_{t'}] \tag{3.9}$$

where $E[X_{t'}]$ is the mean service time of a type t' request including the interruption periods during the service.

3.3.1.5   Mean Residual Times

The mean residual time, $E[\varphi_{busy,t}]$, of a type t busy period is given by [7]

$$E[\varphi_{busy,t}] = \frac{E[T_{busy,t}^2 | N_{y,0} \geq 1]}{2 E[T_{busy,t} | N_{y,0} \geq 1]} \tag{3.10}$$

where

$$E[T_{busy,t}^2] = E[(\sum_{k=0}^{\infty} T_k)^2] = \frac{(1 + b_t)}{(1 - b_t)} E[\sum_{k=0}^{\infty} T_k^2] \tag{3.11}$$

The residual time of a type t busy period exists only if a type t busy period is generated. Therefore, the first and second moments of $T_{busy,t}$ are conditioned on $N_{y,0} \geq 1$. With further manipulation, it can be shown that

$$(1 - b_t^2) \sum_{k=0}^{\infty} E[T_k^2] = E[T_0^2] \tag{3.12}$$

$$+ \frac{E[T_0]}{(1 - b_t)} \lambda_t (\frac{1}{p_1^2} E[Y_t^2] + \lambda_I E[I^2] E[Y_t]).$$

$$E[T_{busy,t}{}^2|N_{y,0}\geq 1] = \frac{E[T_0{}^2|N_{y,0}\geq 1]}{(1-b_t)^2} \tag{3.13}$$

$$+ \frac{E[T_0|N_{y,0}\geq 1]}{(1-b_t)^3}\lambda_t\left\{\frac{1}{p_1{}^2}E[Y_t{}^2]+\lambda_I E[I^2]E[Y_t]\right\}$$

$$E[T_{busy,t}|N_{y,0}\geq 1] = \frac{E[T_0|N_{y,0}\geq 1]}{(1-b_t)} \tag{3.14}$$

where

$$E[T_0{}^2|N_{y,0}\geq 1] = \frac{1}{p_1{}^2}E[T_0{}'^2|N_{y,0}\geq 1]+\lambda_I E[I^2]E[T_0{}'|N_{y,0}\geq 1] \tag{3.15}$$

$$E[T_0|N_{y,0}\geq 1] = \frac{1}{p_1}E[T_0{}'|N_{y,0}\geq 1]. \tag{3.16}$$

As a type t busy period may be one of the three busy period types (type t_a, type t_b, and type t_c), the mean STR of the initiating type t request, $E[T_0{}'|N_{y,0}\geq 1]$ is given by

$$E[T_0{}'|N_{y,0}\geq 1] = \frac{1}{P_{I,t}}\left\{P_{idle,av}\int_0^t v f_\tau(v)dv \right. \tag{3.17}$$

$$+ P_{idle,int}E[N_b|N_b\geq 1]\int_0^t v f_\tau(v)dv/F_\tau(t)$$

$$\left. + P_{busy,t'}E[N_c|N_c\geq 1]\int_t^\infty v f_\tau(v)dv/(1-F_\tau(t))\right\}$$

where N_b and N_c are the numbers of type t requests at the initiation instant of type t_b and type t_c busy periods, respectively. Similar to (3.17), $E[T_0{}'^2]$ is given by

$$E[T_0{}'^2|N_{y,0}\geq 1] = \frac{1}{P_{I,t}}\left\{P_{idle,av}\int_0^t v^2 f_\tau(v)dv \right.$$

$$+ P_{idle,int}\left(E[N_b|N_b\geq 1]\int_0^t v^2 f_\tau(v)dv/F_\tau(t)\right.$$

$$\left. + \left(E[N_b{}^2 - N_b|N_b\geq 1]\int_0^t v f_\tau(v)dv/F_\tau(t)\right)^2\right) \tag{3.18}$$

$$+ P_{busy,t'}\left(E[N_c|N_c\geq 1]\int_t^\infty v^2 f_\tau(v)dv/(1-F_\tau(t))\right.$$

$$\left.\left. + \left(E[N_c{}^2 - N_c|N_c\geq 1]\int_t^\infty v f_\tau(v)dv/(1-F_\tau(t))\right)^2\right)\right\}.$$

In (3.4), the mean residual time, $E[\varphi_{idle,int}]$, of an idle interrupted period is
given by $E[\varphi_{idle,int}] = E[I^2]/2E[I]$. The mean residual time, $E[\varphi_{t'}]$, of a type
t′ request depends on the service time of a type t′ request (which includes
the interruption periods during the service). The service time $X_{t'}$ is given by
$X_{t'} = Y_{t'} + \sum_{j=0}^{N_{I,0}} I_j$, where $Y_{t'}$ is the original STR of a type t′ request, $N_{I,0}$ is
the number of interruptions occurred during T_0' $(=Y_{t'})$, and I_j is the duration of the
j^{th} $(\in \{0, 1, \ldots\})$ interruption. The mean and the second moment of $X_{t'}$ is given by

$$E[X_{t'}] = \frac{1}{p_1} \int_t^\infty v f_\tau(v) dv \qquad (3.19)$$

$$E[X_{t'}^2] = \frac{1}{p_1^2} \int_t^\infty v^2 f_\tau(v) dv + \lambda_I E[I^2] \int_t^\infty v f_\tau(v) dv. \qquad (3.20)$$

Similarly, the mean service time of a type t request is given by

$$E[X_t] = \frac{t}{p_1} \qquad (3.21)$$

As there is no preemption, the mean service time does not depend on the
arrivals, and it only depends on the STR and the mean channel availability. The
probabilities $P_{busy,t}$, $P_{busy,t'}$, and $P_{idle,int}$ are given by $P_{busy,t} = \lambda \int_0^t v f_\tau(v) dv / p_1$,
$P_{busy,t'} = \lambda \int_t^\infty v f_\tau(v) dv / p_1$, and $P_{idle,int} = p_0(p_1 - \lambda \int_0^\infty v f_\tau(v) dv)/p_1$, where
fractions (proportions) of time having a type t and type t′ request occupying
the channel are given by $\lambda \int_0^t v f_\tau(v) dv$ and $\lambda \int_t^\infty v f_\tau(v) dv$, respectively.
The mean residual time, $E[\varphi_{t'}]$, can be evaluated from the standard equation
$E[\varphi_{t'}] = E[X_{t'}^2]/2E[X_{t'}]$.

3.3.2 SPT Service Discipline With Preemption

The SPTWP differs from the SPTNP in that it preempts a current user to give
priority to a new request with an original SRT smaller than that of the current
user. Therefore, a target request with STR equal to t preempts an ongoing type
t′ request to get the channel access, and initiates a type t busy period. From
the viewpoint of a type t user, all the interruptions of a type t idle period
are idle interruptions and all the available durations of a type t idle period are
idle available periods. Therefore, the channel available and interrupted periods of a
type t idle period are denoted as type t idle available (type t_a idle) and
type t idle interrupted (type t_b idle) periods, respectively, as illustrated in
Fig. 3.6.

    Similar to the SPTNP, type t_a and type t_b busy periods are initiated by a
type t arrival in type t_a and type t_b idle periods, respectively. However,
with the SPTWP, there is no type t_c busy period. A target request with STR

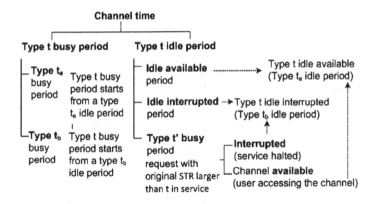

Fig. 3.6 Categorization of channel time for the SPTWP service discipline

equal to t arriving in a `type t` idle available or `type t` idle interrupted period starts its service similar to that arrives in an idle available or idle interrupted period with the SPTNP service discipline, respectively. Examples for initiations of `type` t_a and `type` t_b busy periods are illustrated in Fig. 3.7. A summary of waiting

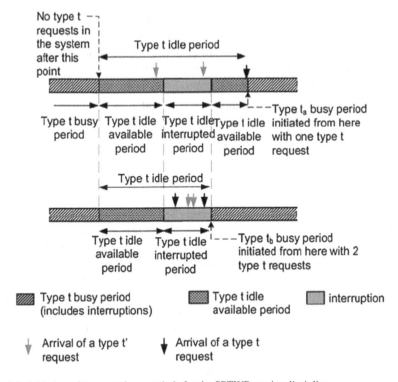

Fig. 3.7 Initiation of `type t` busy periods for the SPTWP service discipline

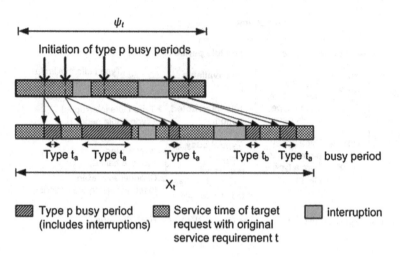

Fig. 3.8 Service time of the target request for the SPTWP service discipline

times of a target request falling into different periods is given in Table 3.1. If the target request arrives in a type t idle available period, it gets the channel access immediately. Therefore, the response time, T_R, is the service time, X_t. However, the service time, X_t, is different from that of the SPTNP, since any type t request arrival during the service time of the target request can preempt the target request. The preempted durations are the durations of type t busy periods initiated during the original service time of the target request. An illustration of the service time X_t is given in Fig. 3.8, where $\psi_t = t + \sum_{j=0}^{N_{I,0}} I_j$, $N_{I,0}$ is the number of interruptions occurred during $T_0' = t$ (in this example $N_{I,0} = 2$), and I_j is the duration of the jth interruption. Analysis of $E[X_t]$ is similar to that of $E[T_{busy,t}]$ given in (3.6) and (3.7) with $E[Y_t]$ equivalent to t and $E[N_{y,0}] = 1$. Therefore,

$$E[X_t] = \frac{t}{p_1(1 - b_t)}. \tag{3.22}$$

When the preemptions are allowed, the server (BS) stops serving a request with STR equal to t (current request) due to an arrival of a type t request. Therefore, the current request does not receive service during a type t busy period. Different from that of SPTNP the mean service time $E[X_t]$ depends on the fraction of time that a type t busy period is in service. Similar to a target request arrival in an idle interrupted period with the SPTNP, a target request arrival in a type t idle interrupted period with the SPTWP has $E[W_t] = E[\varphi_{idle,int,t}] + E[T_{busy,t_b}]P(N_b \geq 0)$, where $E[\varphi_{idle,int,t}]$ $(= E[\varphi_{idle,int}])$ is the mean residual time of the type t idle interruption period.

If a target request arrives in a type t busy period, the mean waiting time is given by $E[W_t] = E[\varphi_{busy,t}]$. The conditional MRT, $E[T_R | \tau = t]$, is given by

$$E[T_R|\tau = t] = \left(E[\varphi_{idle,int}] + E[T_{busy,t_b}]P(N_b \geq 0)\right)P_{idle,int,t}$$
$$+ E[\varphi_{busy,t}]P_{busy,t} + E[X_t] \tag{3.23}$$

where $P_{idle,int,t}$ is the probability of the target request arriving in a `type` t idle interrupted period, $E[\varphi_{idle,int}] = E[I^2]/2E[I]$, and $E[\varphi_{busy,t}]$ is given in (3.8).

A target request can arrive in either of the `type` t busy periods (`type` t_a or `type` t_b). Therefore, $E[\varphi_{busy,t}]$ can be evaluated using (3.6), (3.7), (3.10)–(3.16) with

$$E[T_0'|N_{y,0}\geq 1] = \frac{1}{P_{I,t}}\left\{P_{idle,av,t}\int_0^t vf_\tau(v)dv \right.$$
$$\left. + P_{idle,int,t}E[N_b|N_b \geq 1]\int_0^t vf_\tau(v)dv/F_\tau(t)\right\} \tag{3.24}$$

$$E[T_0'^2|N_{y,0}\geq 1] = \frac{1}{P_{I,t}}\left\{P_{idle,av,t}\int_0^t v^2 f_\tau(v)dv \right.$$
$$+ P_{idle,int,t}\left(E[N_b|N_b \geq 1]\int_0^t v^2 f_\tau(v)dv/F_\tau(t)\right.$$
$$\left.\left. + \left(E[N_b^2 - N_b|N_b \geq 1]\int_0^t vf_\tau(v)dv/F_\tau(t)\right)^2\right)\right\} \tag{3.25}$$

where $P_{I,t} = (1 - P_{busy,t})F_\tau(t)$ is the probability of a request arrival initiating a `type` t busy period and $P_{idle,av,t}$ is the probability of the target request arriving in a `type` t idle available period. Equations (3.24) and (3.25) differ from (3.17) and (3.18) in that (3.24) and (3.25) do not contain the components for a `type` t_c busy period. The probabilities $P_{busy,t}$, $P_{idle,av,t}$, and $P_{idle,int,t}$ are given by, $\lambda \int_0^t vf_\tau(v)dv/p_1$, $p_1 - \lambda \int_0^t vf_\tau(v)dv$, and $p_0(p_1 - \lambda \int_0^t vf_\tau(v)dv)/p_1$, respectively.

3.3.3   SRPT Service Discipline

The SRPT differs from the SPTWP in that it compares the remaining STRs of the service requests rather than their original STRs. Therefore, a `type` t' request always initiates a `type` t busy period when its remaining STR reduces to t, and an incoming request with STR equal to t can preempt a `type` t' request only when the remaining STR of the `type` t' request is larger than t. In order to capture the difference, we alter the definition of `type` t busy period as a continuous time period during which services with the **remaining** STR **less** than t are using or being interrupted while using the channel. Similarly, the definition of `type` t' busy period is altered as a continuous time period during which `type` t' requests with the **remaining** STR **greater** than t are using or being interrupted while

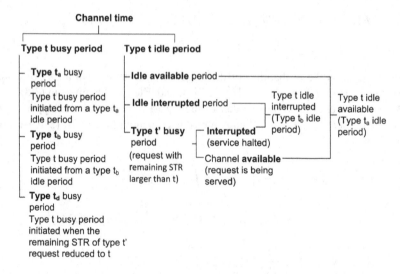

Fig. 3.9 Categorization of channel time for the SRPT service discipline

using the channel. The categorization of the time periods is illustrated in Fig. 3.9. As illustrated in Fig. 3.9, a type t busy period may be any of type t_a, type t_b, and type t_d busy periods. Similar to that of the SPTWP service discipline, the type t_a and type t_b busy periods are initiated due to type t arrivals during type t_a and type t_b idle periods, respectively. However, a type t_d busy period is initiated when the remaining STR of a type t' request becomes t. The waiting times of a target request with the original STR up to t are given in Table 3.1. Similar to that of the SPTWP, the expression for the MRT of a target request is given by

$$E[T_R|\tau = t] = \left(E[\varphi_{idle,int,t}] + E[T_{busy,t_b}]P(N_b \geq 0)\right) P_{idle,int,t} \quad (3.26)$$
$$+ E[\varphi_{busy,t}]P_{busy,t} + E[X_t]$$

where $E[\varphi_{idle,int,t}]$, $E[T_{busy,t_b}]$, and N_b are the same as those with the SPTWP. The probabilities $P_{busy,t}$ and $P_{idle,int,t}$ are given by $\lambda \left[\int_0^t v f_\tau(v)dv + t(1 - F_\tau(t)) \right]$ $/p_1$, $p_0 \left(1 - P_{busy,t}\right)$, respectively, where the numerator in $P_{busy,t}$ is the fraction of time that requests with the remaining STR less than t occupies the channel (excluding the interruption durations). The evaluation of $E[\varphi_{busy,t}]$ in (3.26) is similar to that given in (3.17) with

$$E[T_0'|N_{y,0}\geq 1] = \frac{1}{P_{I,t}}\left\{P_{idle,av,t} \int_0^t v f_\tau(v)dv + t[1 - F_\tau(t)] \right. \quad (3.27)$$
$$\left. + P_{idle,int,t} E[N_b|N_b \geq 1] \int_0^t v f_\tau(v)dv/F_\tau(t)\right\}$$

$$E[T_0'^2|N_{y,0}\geq 1] = \frac{1}{P_{I,t}}\left\{ P_{idle,av,t}\int_0^t v^2 f_\tau(v)dv + t^2\left(1 - F_\tau(t)\right)\right. \tag{3.28}$$

$$+ P_{idle,int,t}\left(E[N_b|N_b \geq 1]\int_0^t v^2 f_\tau(v)dv/F_\tau(t)\right)$$

$$\left. + P_{idle,int,t}\left(E[N_b^2 - N_b|N_b \geq 1]\int_0^t vf_\tau(v)dv/F_\tau(t)\right)^2\right\}$$

where $P_{I,t}$ is the probability of an incoming request initiating a type t (type t_a, type t_b, or type t_d) busy period. Therefore, $P_{I,t}=P_{idle,av,t}F_\tau(t)$ $+P_{idle,int,t}F_\tau(t)+1 - F_\tau(t)= 1 - P_{busy,t}F_\tau(t)$.

In the case of SRPT, a new service request can preempt the current request only if the STR of the new request is less than the remaining STR of the current request at the arrival instant. Therefore, the service time comparison has to be done exactly at the arrival instant of the new request. This comparison is not possible in continuous-time as the probability that an arrival occurs at a particular time instance is zero. It is only possible to find the probability of request arrivals with the original STR shorter than the remaining service time of the current request for a given period of time. As a result, we divide the service time requirement t (or equivalently the file length) of the target request into n units of duration Δt ($t = n\cdot\Delta t$) as illustrated in Fig. 3.10, where a type $(n - i)\Delta t$ busy period is similar to a type t busy period which starts from a type $(n - i)\Delta t$ request and ends after serving all such requests in the waiting queue, and a type $(n - i)\Delta t$ request being a service request with the original STR less than $(n - i)\Delta t$. The service time $X_{\Delta t,i}$ ($i \in \{1, 2, \ldots, n\}$) is the actualduration it takes to complete the ith unit of Δt, and $X_t=\sum_{i=1}^{n} X_{\Delta t,i}$ [16].

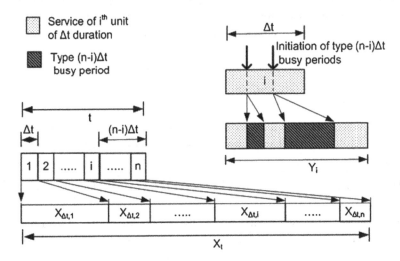

Fig. 3.10 Service time for the SRPT service discipline

We have $E[X_t]=\sum_{i=1}^{n} E[X_{\Delta t,i}]$. The duration Δt in $X_{\Delta t,i}$ is equivalent to T_0' in $T_{busy,t}$. Therefore, similar to the evaluation of $E[T_{busy,t}]$, $E[X_{\Delta t,i}]$ is given by $E[X_{\Delta t,i}] = \lim_{m\to\infty}\sum_{k=0}^{m+1} T_{i,k}$ with

$$E[T_{i,k+1}|T_{i,k}] = \frac{1}{p_1}(\lambda_{(n-i)\Delta t}T_{i,k})\frac{1}{f_\tau((n-i)\Delta t)}\int_0^{(n-i)\Delta t} v f_\tau(v)dv \qquad (3.29)$$

where $\lambda_{(n-i)\Delta t}T_{i,k}$ represents the mean number of request arrivals with the STR less than $(n-i)\Delta t$ during T_k, and the integral represents the mean STR of such arrival. Similar to $E[T_{busy,t}]$, the mean duration $E[X_{\Delta t,i}]=E[T_0]/(1-b_{(n-i)\Delta t})$, where $E[T_0]=\Delta t/p_1$ and $b_{(n-i)\Delta t}=\lambda_{(n-i)\Delta t}\int_0^{(n-i)\Delta t} v f_\tau(v)dv/p_1$, $F_\tau((n-i)\Delta t)$. In order to make the analysis accurate (close to that for the continuous-time scenario), Δt has to be very small (i.e., n is very large). As our original system is time-slotted, we can carry out the analysis in discrete-time with $\Delta t = 1$ time unit, which is the size of a time-slot. However, since our analysis so far has been in continuous-time, we divide the time into infinitely small (i.e., a large number of) time periods. Therefore, the service time, $E[X_t]$, is obtained by making Δt very small ($\Delta t \to 0$) or equivalently n very large ($n\to\infty$) [2].

Duration of type t busy periods are independent and identically distributed with SPTWP and SRPT service disciplines, and the inter-arrival time of the new requests are memory-less. In the analysis of residual time of type t busy period given that the target request arrives in a type t busy period, we can ignore the type t idle periods, and consider the initiation of a type t busy period as a renewal process. However, durations of type t' services and type t busy periods can be weakly dependent in the case of SPTNP service discipline. The dependence vanishes with the occurrence of an idle available period (when the server becomes idle). Therefore, we assume these durations to be independent and evaluate the residual times similar to the cases of SPTWP and SRPT.

3.3.4 PS Service Discipline

The conditional MRT of a target request operating over a network following exponentially distributed channel availability durations is given by [17]

$$E[T_R|\tau = t] = \frac{t}{p_1(1-\sigma)} + p_0\frac{E[I^2]}{2E[I]} + p_0\frac{\sigma E[I^2]}{2E[I]}\frac{2-\sigma}{(1-\sigma)^2}(1-e^{-\frac{(1-\sigma)t}{E[\tau]}}) \qquad (3.30)$$

where $\sigma = \lambda E[\tau]/p_1$ is the utilization factor (ratio between the mean arrival rate and mean service rate).

---

[2]In our analysis, we set $n = 10^4$.

3.4 Comparison of the Service Disciplines

For presentation clarity, we use subscripts NP, WP, and SRPT for the components associated with SPTNP, SPTWP, and SRPT service disciplines, respectively. The difference in the conditional MRTs between SPTNP and SPTWP is given by

$$E[T_R|\tau = t]_{NP} - E[T_R|\tau = t]_{WP} = (\xi_{t'} - p_0\xi_{t_b})(\sigma - b_t) \qquad (3.31)$$

$$+ b_t(E[\varphi_{busy,t}]_{NP} - E[\varphi_{busy,t}]_{WP}) - \frac{t}{p_1} \cdot \frac{b_t}{(1 - b_t)}$$

where $\xi_{t'} = E[\varphi_{t'}] + E[T_{busy,t_c}]P(N_c \geq 1)$, $\xi_{t_b} = E[\varphi_{idle,int}] + E[T_{busy,t_b}]P(N_b \geq 1)$, and $\frac{t}{p_1} \cdot \frac{b_t}{(1-b_t)}$ is the service time difference between SPTNP and SPTWP for a data file requiring a service time equal to t. The terms $(\xi_{t'} - p_0\xi_{t_b})$ and $(E[\varphi_{busy,t}]_{NP} - E[\varphi_{busy,t}]_{WP})$ contain busy periods initiated from more than one request arrival, whereas $\frac{t}{p_1} \cdot \frac{b_t}{(1-b_t)}$ only contains the service time of a target request. The probability b_t monotonically increases with t. The terms $(\xi_{t'} - p_0\xi_{t_b})$, $(\sigma - b_t)$, and $E[\varphi_{busy,t}]_{NP} - E[\varphi_{busy,t}]_{WP}$ decreases with t, and $\frac{t}{p_1} \cdot \frac{b_t}{(1-b_t)}$ increases with t. Therefore, the difference in the conditional MRTs $E[T_R|\tau = t]_{NP} - E[T_R|\tau = t]_{WP}$ varies from a very high positive value to a small negative value as t increases. When the file size Weibull (heavy tail) distributed as in (3.1), the probability of having a smaller file size is high and that of a larger file size is low. Therefore, the resultant MRT difference $E[T_R]_{NP} - E[T_R]_{WP}$ obtained by averaging $E[T_R|\tau = t]_{NP} - E[T_R|\tau = t]_{WP}$ over t is a positive value. The probability of having very large and very small values for $E[T_R|\tau = t]_{NP} - E[T_R|\tau = t]_{WP}$ increases with the tail heaviness in the file size distribution. As a result, $E[T_R]_{NP} - E[T_R]_{WP}$ increases with the tail heaviness in the file size distribution. Increment of the terms $(\xi_{t'} - p_0\xi_{t_b})$ and $(E[\varphi_{busy,t}]_{NP} - E[\varphi_{busy,t}]_{WP})$ with an increasing interruption duration is larger than that of $\frac{t}{p_1} \cdot \frac{b_t}{(1-b_t)}$. Therefore, the MRT difference $E[T_R]_{NP} - E[T_R]_{WP}$ increases with the interruption duration. The difference between the conditional MRTs for SPTWP and SRPT is given by

$$E[T_R|\tau = t]_{WP} - E[T_R|\tau = t]_{SRPT} = E[X_t]_{WP} - E[X_t]_{SRPT} \qquad (3.32)$$

$$+ b_t^* p_1 E[T_{t_b}] - b_t(E[\varphi_{busy,t}]_{SRPT} - E[\varphi_{busy,t}]_{WP})$$

$$- b_t^*(E[\varphi_{busy,t}]_{SRPT} - p_1 E[\varphi_{idle,int}])$$

where $b_t^* = t(1 - F_\tau(t))/p_1$. The mean service time $E[X_t]_{WP}$ is greater than $E[X_t]_{SRPT}$, due to the larger number of preemptions in SPTWP than SRPT, and the difference $E[X_t]_{WP} - E[X_t]_{SRPT}$ increases with t. The terms $E[T_{t_b}]$ and $(E[\varphi_{busy,t}]_{SRPT} - E[\varphi_{busy,t}]_{WP})$ are smaller positive values and $(E[\varphi_{busy,t}]_{SRPT} - p_1 E[\varphi_{idle,int}])$ is negative for a smaller t value. However, all three terms are larger positive values for a larger t. Therefore, difference in conditional MRTs $E[T_R|\tau = t]_{WP} - E[T_R|\tau = t]_{SRPT}$ varies from a small positive value to a small negative value with increasing t. Similar to (3.31), the MRT difference $E[T_R]_{WP} - E[T_R]_{SRPT}$ is a positive value when the file lengths are heavy tail distributed. However, this positive value is smaller than that in (3.31). The difference between the conditional MRTs for SPTNP and SRPT is given by

$$E[T_R|\tau = t]_{NP} - E[T_R|\tau = t]_{SRPT} = (\xi_{t'} - p_0\xi_{t_b})(\sigma - b_t) \qquad (3.33)$$
$$+ b_t(E[\varphi_{busy,t}]_{NP} - E[\varphi_{busy,t}]_{SRPT}) + E[X_t]_{NP} - E[X_t]_{SRPT}.$$

Similar to the discussion on (3.31), the terms $(\xi_{t'} - p_0\xi_{t_b})$ and $(E[\varphi_{busy,t}]_{NP} - E[\varphi_{busy,t}]_{SRPT})$ contain busy periods initiated using one or more request arrival, whereas $E[X_t]_{SRPT} - E[X_t]_{NP}$ only contains the service time of a target request. Further, the terms $(\xi_{t'}-p_0\xi_{t_b})$, $(\sigma - b_t)$, and $E[\varphi_{busy,t}]_{NP} - E[\varphi_{busy,t}]_{SRPT}$ decrease, and $E[X_t]_{SRPT} - E[X_t]_{NP}$ increases with t. Therefore, the difference in conditional MRTs $E[T_R|\tau = t]_{NP} - E[T_R|\tau = t]_{SRPT}$ varies from a very high positive value to a very small value with increasing t, and the unconditional MRT difference $E[T_R]_{NP} - E[T_R]_{SRPT}$ is a positive value when the file length is heavy tail distributed. Further, the difference $E[T_R]_{NP} - E[T_R]_{SRPT}$, increases with the interruption duration and the tail heaviness of the file length distribution.

3.5   Numerical Results

Computer simulations are carried out to evaluate the accuracy of the response time analysis. As the system is time-slotted, the simulations are in discrete time and the time is measured in time-slot units. Therefore, the STR of a service request is measured in number of time-slots. The BS transmits packets to the SUs in idle time-slots (which are not being used by the PUs) based on four service disciplines, respectively. The BS transmits only one packet in each idle time-slot. Service requests are generated according to a Poisson arrival process with a Weibull distributed file length. The MRT, $E[T_R]$, is evaluated by averaging the results of 20 simulation runs, each run having 18,000 service requests. Note that, the mean channel available and unavailable (interruption) durations are denoted as T_{on} and T_{off} throughout this section.

   Figure 3.11 shows the variation of $E[T_R]$ with T_{on} and T_{off} obtained from numerical analysis and simulations while having $T_{off} = 10$ and $T_{on} = 10$ time-slots, respectively for all four service disciplines in a light traffic load condition. We keep $\sigma = 0.6$ and $\alpha = 0.6$. The results demonstrate that the simulation results closely match with the analytical results for all four service disciplines. When preemption is allowed, the MRT decreases considerably, and using the remaining STR instead of the original STR improves the performance. The PS outperforms the SPTNP service discipline for the lightly loaded system. The MRT decreases exponentially with the channel availability and increases with the mean interruption duration.

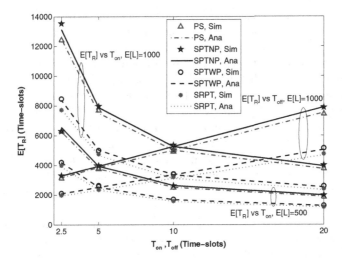

Fig. 3.11 The variation of MRT with mean channel availability and interruption durations at $\sigma = 0.6$

Figure 3.12 shows the variation of $E[T_R]$ with σ obtained from numerical analysis and simulations for all four service disciplines, with $T_{on} = 20$, $T_{off} = 10$ time-slots, $E[L] = 500$, and $\alpha = 0.6$. It is observed that the response times of all

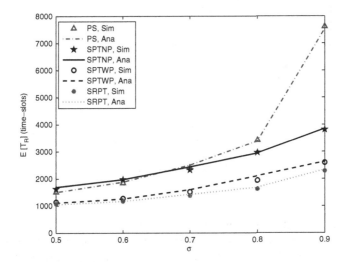

Fig. 3.12 The variation of MRT with the traffic load

four service disciplines increase with σ, and the larger the σ, the larger the rate of increment of $E[T_R]$. As the mean service rate remains constant, the mean arrival rate is proportional to the offered traffic load, and the larger the arrival rate, the longer the waiting time of the users at the waiting queue. Therefore, the waiting time increases with the traffic load at the BS, leading to longer response times. As seen in Fig. 3.11, the service disciplines with preemption outperforms that without preemption, and the PS outperforms the SPTNP in a lightly loaded condition. For the PS service discipline, the heavier the traffic load, the larger the number of users in the round-robin order. Therefore, the mean service time increases for each request; whereas for the SPTNP service discipline, the increasing number of requests has a major impact on the waiting time (or the response time) of the requests with a long STR, and vice versa. However, the probability of request arrivals with a long STR is small. Therefore, the rate of increment of the MRT with the traffic load is larger for the PS service discipline than that for the SPTNP service discipline. This rapid increment is indeed captured in (3.30).

Figure 3.13 shows the $E[T_R]$ variation with T_{off} for the SRPT service discipline obtained from numerical analysis and simulations for two different traffic load conditions and α values with $T_{on} = 100$ time-slots and $E[L] = 500$.

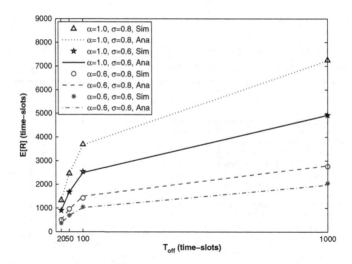

Fig. 3.13 The variation of MRT with the mean interruption duration for the SRPT service discipline

The simulation results closely match with the analytical results. That is, the discrete-time analysis in Sect. 3.3.3 is accurate for the networking scenario. Similar to what is observed in Fig. 3.12, the $E[T_R]$ increases with σ. Further, the heavier the tail of the STR distribution, the shorter the MRT.

Figure 3.14 shows the $E[T_R]$ variation with the tail heaviness (shape parameter) of the file length (or the STR) with $\sigma = 0.6$, $T_{on} = 20$, and $T_{off} = 10$ time slots. The MRT of the PS service discipline remains almost the same with the variation of α. As the PS gives an equal opportunity to all service requests, the $E[T_R]$ depends

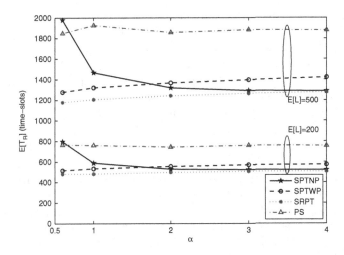

Fig. 3.14 The variation of MRT with the tail heaviness of the STR distribution

on the mean of the STR, not its distribution [18]. When the preemption is allowed, the MRT decreases with the heaviness of the tail; otherwise, it increases with the heaviness of the tail. Preemptions result in shorter response times for requests with a short STR and longer response times for requests with a long STR. The smaller the α, the larger the number of service requests with very short STR. Therefore, when the preemption is allowed, the heavier the tail of the STR distribution (a lower α), the shorter the $E[T_R]$. When α is large, the STR concentrate around its mean, and the probability of an incoming request having a shorter original STR than the remaining STR of the current user is very low. This reduces the probability of having preemptions by a large margin in the case of the SRPT. When the preemptions are not likely to happen, the SRPT is similar to the SPTNP. Therefore, the larger the α, the closer the $E[T_R]$ of the SPTNP to that of the SRPT. Incase of the SPTWP, the preemptions are carried out with respect to the original STR of both the incoming and existing requests. When the STR concentrate around its mean, the remaining STR of most of the ongoing services are less than the original STR of the incoming service requests. Therefore, the probability of the incoming service request having a shorter original STR than that of the ongoing service is larger than the probability of incoming service request having a shorter original STR than the remaining STR of the ongoing service. As a result, the number of preemptions are larger with the SPTWP than that with the SRPT. During most of the preemptions, the ongoing services with shorter remaining STR are preempted by the service requests with long original STR. Therefore, the preemotions increase the waiting times, leading to longer MRTs in the case of the SPTWP than that in the case with the SPTNP and SRPT. Figure 3.14 demonstrates that the heavier the tail of the STR distribution, the longer the MRT, and the larger the rate of increment of the mean STR for the

SPTNP service discipline. Therefore, the SPTNP service discipline is not a good choice when the tail of the STR distribution is very heavy.

Figure 3.15 shows the $E[T_R]$ variation with T_{off} at $p_1 = T_{on}/(T_{on}+T_{off})$=0.66 with $E[L] = 500$ and $\sigma = 0.6$ for two different α values. The $E[T_R]$ increases with T_{off} even when the mean (long term) channel availability and the traffic load remain unchanged. When the interruption duration is exponentially distributed, the conditional MRT for the SPTWP service discipline given in (3.23) can be simplified to $E[T_R|\tau = t]=p_0 T_{off}+E[\varphi_{busy,t}]b+\frac{t}{p_1(1-b_t)}$. Note that $E[\varphi_{busy,t}]$ increases and b_t remains constant with T_{off} for constant p_1 and σ. As a result, the longer the T_{off} the longer the $E[T_R|\tau = t]$. Similarly, we can show that the conditional MRTs of the SPTNP, SRPT, and PS service disciplines increase with the mean interruption duration, when the long term channel availability and the traffic load remain unchanged. Similar to Fig. 3.14, the shorter the tail of the STR distribution (larger α), the shorter the $E[T_R]$ for the SPTNP and the longer the $E[T_R]$ for the rest of the service disciplines. As in (3.31) and (3.33), the difference between the MRTs of the SPTNP and SPTWP and that between the SPTNP and SRPT increase with T_{off}. However, there is no significant difference in the gap between the MRTs of the SPTWP and SRPT with the variation of T_{off}.

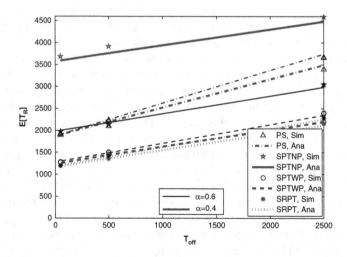

Fig. 3.15 The variation of MRT with the mean channel unavailability duration for constant p_1

Figure 3.16 shows the CDFs of the file length L (or STR) and the response times obtained from simulations for the SPTNP and SPTWP service disciplines, respectively, with $E[L] = 500$, $T_{on} = 10$ and $T_{off} = 20$ time-slots, and $\sigma = 0.6$. It is observed that a higher percentage of requests show a very short response times when the preemption is allowed. Further, the slope of the response time curve reduces for a high STR (file length) value with the SPTWP than that with the SPTNP. This means that the response time for a long STR is getting longer when the preemption is allowed. Preemption gives a higher service priority to requests

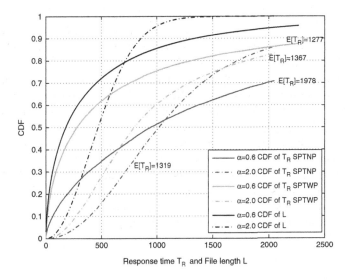

Fig. 3.16 The CDF of MRTs of both SPT service disciplines and the STR

with a short STR over requests with a long STR. Therefore, the requests with a short STR experience a very short response time and the requests with a long STR experience a very long response time. That is, the preemption compromises the performance of requests with a long STR to other requests with a short STR. This may not be fair in the viewpoint of the requests with a long STR. If we try to give a higher priority to the requests with a long STR over the requests with a short STR, the latter will have to wait for a very long time before getting the service, and the resulting MRT will be longer. However, the requests with a long STR will get a shorter response time than that in the case with SPR and SRPT disciplines. Based on our analysis, the MRT can be evaluated for a system, given the channel availability statistics, STR (file length) distribution, and request arrival rate. The best service discipline can then be selected based on the MRT requirement of the data service and desired trade-off with service fairness.

In this work, the MRT is studied with respect to a time-slotted cognitive radio network. As described in Sect. 3.3, we conduct the analysis in continuous time, considering the fact that the interested time durations (STR, T_{on}, and T_{off}) are larger when compared with the time-slot duration. Given that an accurate channel sensing can be done, this analysis can be directly applied to the continuous-time scenario (i.e., when the channel is not time-slotted). Further, in this work, we consider only the single channel scenario. In a multiple channel network, two key approaches can be considered:

1. The BS divide incoming requests among the channels based on the arrival sequence, and SUs stay in the assigned channel for the service;
2. The BS assigns a channel to SUs instantaneously based on the channel availability in each time-slot.

The two approaches differ in signal/channel-switching overhead and in statistical multiplexing performance gain. The model here can be applied to the first scenario where the arrival rate should be normalized to the number of channels in the network. Extension to the second scenario requires further research. This analysis can be used as a benchmark for the performance in the second scenario. Further, this work is focused on the operation of a CRN with single base station. In the case of multiple base stations, data call handover between neighboring base stations should be considered. The file length (or equivalently the STR) distribution and the arrival process for each base station is a combination of those of the new request arrivals and the handover data calls. This can be treated in a way similar to that in [4, 8, 19] in which the authors consider handover between a cellular network and a WLAN. In extending the analysis to a system with multiple base stations, different file size distributions and different arrival processes (for new and handover calls) should be considered, which is expected to be much more complex.

3.6   Summary

In this chapter, we have evaluated the MRT of elastic data traffic under three service disciplines (shortest processor time without preemption, shortest processor time with preemption, and shortest remaining processing time) for a single-channel single-hop synchronized CRN with a base station, in comparison with the PS service discipline. Numerical results demonstrate that the SRPT and the SPTWP provide very close response times and the SRPT outperforms the SPTWP. The MRTs for all four service disciplines are compared under different traffic load conditions, and it is shown that the SPTNP service discipline outperforms the PS service discipline in heavy traffic load conditions. Therefore, the SPTNP service discipline is a better choice over the PS service discipline as the traffic load increases. The MRTs of all four service disciplines are compared under the Weibull distribution with different tail parameters, and the results show that the preemption reduces the MRT when the service time requirement follows a heavy tailed distribution. The SRPT performs better than the other service disciplines in terms of MRT, as it achieves very short response times for service requests with short service time requirements. Further, the SPTNP is not a good choice when the tail of the STR distribution is very heavy. This MRT analysis can be used to choose the best service discipline based on the given system parameters and the service quality requirement of the users. Further, the mean duration of the transmission interruptions (channel non-available durations) has an impact on the MRT even when the long term channel availability and the offered traffic load at the BS remain unchanged. The MRT analysis can be used for call admission control to ensure service satisfaction.

References

1. Li, H.: Impact of primary user interruptions on data traffic in cognitive radio networks phantom jam on highway. In: Proc. IEEE GLOBECOM (2011)
2. Elias, J., Martignon, F.: Integrated admission control for streaming and elastic traffic. In: Proc. IEEE ICC (2010)
3. Massoulíe, L., Roberts, J.W.: Bandwidth sharing and admission control for elastic traffic. Telecommunication systems 15(1–2), 185–201 (2006)
4. Song, W., Zhuang, W.: Multi-class resource management in a cellular/wlan integrated network. In: Proc. IEEE WCNC, pp. 3070–3075 (2007)
5. Lin, M., Wierman, A., Zwart, B.: The average response time in a heavytraffic srpt queue. SIGMETRICS Perform. Eval. Rev. 38(2), 12–14 (2010)
6. Kartheek, M., Misra, R., Sharma, V.: Performance analysis of data and voice connections in a cognitive radio network. In: Proc. IEEE NCC (2011)
7. Avi-Itzhak, B., Naor, P.: Some queuing problems with the service station subject to breakdown. Operations Research 11(3), 303–320 (1963)
8. Song, W., Zhuang, W.: Multi-service load sharing for resource management in the cellular/wlan integrated network. IEEE Trans. Wireless Commun. 8(2), 725–735 (2009)
9. Gunawardena, S., Zhuang, W.: Service response time of elastic data traffic in cognitive radio networks with spt service discipline. In: Proc. IEEE GLOBECOM (2012)
10. Gunawardena, S., Zhuang, W.: Service response time of elastic data traffic in cognitive radio networks. IEEE J. Select. Areas Commun. 31(3), 559–570 (2013)
11. Rezaul, K.M., Pakštas, A.: Web traffic analysis based on edf statistics. In: Proc. 7th Annual Post Graduate Symposium on the Convergence of Telecommunications, Networking and Broadcasting (PGNet) (2006)
12. Wang, P., Niyato, D., Jiang, H.: Voice service capacity analysis for cognitive radio networks. IEEE Trans. Wireless Commun. 59(4), 1779–1790 (2010)
13. Schrage, L.E., Miller, L.W.: The queue m/g/1 with the shortest remaining processor time discipline. Operations Research 14(4), 670–684 (1966)
14. Cobham, A.: Priority assignment in waiting line problems. Operations Research 2(1), 70–76 (1954)
15. Nan, H., Hyon, T., Yoo, S.: Distributed coordinated spectrum sharing mac protocol for cognitive radio. In: Proc. IEEE DySPAN, pp. 240–249 (2007)
16. Hsu, A., Wei, D., Kuo, C.: A cognitive mac protocol using statistical channel allocation for wireless ad-hoc networks. In: Proc. IEEE WCNC, pp. 105–110 (2007)
17. Delcoigne, F., Proutière, A., G.Régnié: Modeling and integration of streaming and data traffic. Perform. Eval. 55(3–4), 185–209 (2004)
18. Jia, J., Zhang, Q., Shen, X.: Hc-mac: A hardware-constrained cognitive mac for efficient spectrum management. IEEE J. Select. Areas Commun. 26(1), 106–117 (2008)
19. Song, W., Zhuang, W.: Resource allocation for conversational, streaming, and interactive services in cellular/wlan interworking. In: Proc. IEEE GLOBECOM, pp. 4785–4789 (2007)

Chapter 4
Service Response Time of Interactive Data Traffic

4.1 Introduction

So far, we have discussed the mean response time of data sessions consisting of single file request (service request). After completing the reception of a data file, the user leaves the system, or places the next service request after a long time compared to the inter-arrival time of the new sessions at the BS. Therefore, successive service requests from a particular user are treated as independent requests. However, in reality, very often each user (such as one browsing the web) places multiple service requests one after another before taking a break from browsing the web, and successive service request arrivals from a particular user are correlated. Therefore, each data session may contain multiple file requests with a thinking (file reading/viewing) time between the completion of one service request to the placement of next service request. For the presentation clarity, we use the term interactive data traffic to denote the data sessions, each with multiple service requests. The response times of interactive data sessions depend on the relationship between the request arrivals of a data session. For service quality provisioning, it is necessary to analyze the response times taking into account the correlation of the successive service requests.

The MRT evaluation of interactive data traffic is studied for cellular/WLAN integrated networks in [1, 2]. In the studies, the network supports streaming and elastic data traffic flows. The SRPT and PS service disciplines are considered in [1] and [2], respectively. The MRT of interactive data traffic over shared packet networks is studied in [3]. The Web traffic is considered as interactive data traffic, and an EDGE (Enhanced Data Rates for GSM Evolution) network is considered as the packet network. In all the existing works, the short-term mean channel rate available for a data user does not vary with time, and therefore the long-term mean channel rate is used for the response time analysis. However, in CRNs, the channel availability for SUs varies with time due to the interruptions by the PUs, and the short-term mean channel availability deviates from the long-term mean

S. Gunawardena and W. Zhuang, *Modeling and Analysis of Voice and Data in Cognitive Radio Networks*, SpringerBriefs in Computer Science, DOI 10.1007/978-3-319-04645-7_4, © The Author(s) 2014

channel availability. Therefore, the effect of the transmission interruptions caused by the PUs should be considered in the analysis. There are limited studies in the literature on the performance analysis of interactive data traffic over CRNs.

4.1.1 Motivation and Objectives

In a cognitive radio network, the service interruptions due to the presence of PUs increase the response time. The number of file requests in an interactive data session can increase the offered traffic load at the BS and increase the response times. Different service disciplines can be used to provide different priorities to the service requests based on their STRs. Therefore, to achieve a proper MRT, it is necessary to study how the MRT changes with networking environment, including channel available statistics, service disciplines, service request arrival process, thinking time statistics, and service time requirement. In this work, we study the MRT of interactive data traffic over CRNs under different service disciplines.

4.1.2 Contributions

The contribution of this chapter is two fold: (i) Given Poisson session arrivals, we show that the data file request arrival process at the BS can be approximated by a Poisson process for all four service disciplines, when the mean channel unavailable duration is short and the offered traffic load is low. Further, the larger the mean number of data file requests per service session, the closer the request arrival process to the Poisson process. However, the request arrival process at the BS deviates from the Poisson process for the SPTNP service discipline when the STR is heavytail distributed; (ii) Under the Poisson approximation, we show that the MRT of a file in an interactive data session can be given by that of elastic data traffic in low and medium traffic load conditions, by applying an equivalent request arrival rates at the BS. However, the MRT approximation does not work well in a heavy traffic load condition.

4.2 System Model

Consider a CRN BS and a number of SUs. The secondary network operates over a time-slotted single-channel primary network. We focus on the downlink information packet transmission from the BS to the SUs. There is a low-rate signaling channel for the SUs to transmit service request packets.

4.2.1 Channel Availability Model

The channel is partitioned into slots of constant duration, and transmission from the PUs are synchronized with the time-slots. The channel is busy if the PUs are using the channel, and idle otherwise. The channel state in each time-slot is denoted as 0 if the channel is busy, and 1 otherwise. The state transition of the channel among adjacent time-slots can be illustrated using a two-state Markov chain, where $S_{i,j}$ denotes the transition probability from state i to state j $(i, j \in \{0, 1\})$ [4, 5]. The channel state can be independent or dependent among adjacent time-slots. The BS can transmit information packets only when the channel is at state 1 (available for SUs), and the channel sensing and information packet transmissions are free of errors. The appearance of a PU transmission at the end of an idle period is denoted as service interruption to the secondary network. The mean channel available and unavailable (interruption) durations are denoted as T_{on} and T_{off}. The channel availability information is consistent throughout the network (spectrum homogeneous).

4.2.2 Interactive Data Traffic

The secondary network supports a large number of interactive data users. Each user starts a data session by placing a service request at the BS. The interactive data session requests arrive at the BS according to a Poisson process with mean arrival rate λ. After receiving a data file, the user reads/views the file and places the next file request at the BS. A user may place multiple such requests one after another before taking a break. Therefore, each data session can be modeled as a collection of a finite number of file transfers and thinking durations (reading/viewing) as illustrated in Fig. 4.1.

Fig. 4.1 Structure of an interactive session. *Shaded* areas represent the file transfers [1]

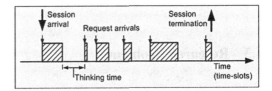

The response time, T_R, is the time duration from the instant that a user places a file request at the BS until it receives the final data packet of the file, and the thinking time, T_T, is the time duration from the user receiving the final packet of a requested file until it places the next service request at the BS. The file transfer and thinking durations are also referred to as on and off phases, respectively. An SU places a file request at the BS via the signaling channel. The number, N_f, of file requests in a data session is geometrically distributed with mean $1/P_T$. We use the terms service request to denote a file request of an SU, user arrival to denote a session arrival at the BS, and completion of a service request to denote the completion of a file transfer. After completion of each service request, an SU decides whether to end the session or to enter the thinking state. If $N_f = 1$, an interactive data session is reduced to an elastic data session. During each available time-slot, only one data user is being served. The size of a data packet transferred during a time-slot is denoted as L_{pk}, which is same for all the time-sots. The duration of an on phase depends on the STR of the request. The STR depends on the length, L, of the requested data file, and L_{pk}. The lengths of requested files are independently and identically distributed with a Weibull distribution which characterizes Internet data traffic [6,7]. Similar to that in Sect. 3.2.3, the PDF, $f_L(\cdot)$, of file length L is given by

$$f_L(x) = \frac{\alpha}{\beta}\left(\frac{x}{\beta}\right)^{\alpha-1} e^{-\left(\frac{x}{\beta}\right)^{\alpha}}, \quad \alpha > 0, \beta > 0, x > 0 \qquad (4.1)$$

where α and β are the shape and scale parameters, respectively. The shape parameter governs the tail heaviness of the Weibull distribution. The file length is heavytail distributed when $\alpha < 1$, and the special case of exponential distribution is given by $\alpha = 1$. Upon a service request, the whole data file is available at the BS without any delay. The MRT of a data file is considered as the QoS parameter of the interactive data traffic, and the QoS requirement is given by $E[T_R] \leq \bar{T}_{R,\max}$, where $\bar{T}_{R,\max}$ is the maximum allowable MRT for satisfactory service quality.

4.2.3   Service Disciplines

The secondary base station serves the secondary data users in shortest processor time without preemption, shortest processor time with preemption, shortest remaining processor time, and processor sharing service disciplines, respectively, as described in Sect. 3.2.4.

4.3   Research Problem

In this work, our objective is to evaluate the MRT of service requests in interactive data sessions, which should be controlled below the desired limit to provide the required service quality. From Chap. 3, it is clear that the MRT of service requests depends on the request arrivals at the BS, STR distribution, and the channel available/unavailable statistics. In order to keep the MRT within the desired limit, it is required to find a relationship between the MRT and networking environment (service discipline, STR distribution, data session arrival process, channel availability statistics, thinking time statistics, and number of service requests per data session). Unlike in the elastic data traffic, in addition to service requests from new session arrival, at the BS, there are request arrivals from the users returning from the thinking state (after reading/viewing the last file). These request arrivals follow a general process, which depends on the statistics of the thinking duration and the service process of the BS. Therefore, is not straight forward to characterize the service request arrival process at the BS. As the first step, we model the system as a queuing network [1] to capture the effects of the BS service process and the thinking process on the request arrivals at the BS. A service request from an SU arrives at the BS, stays at the BS queue[1] until its service completion, and then enters the thinking state. The BS service disciplines (PS, SPTNP, SPTWP, and SRPT) are general processes and the session arrivals follow a Poisson process. Therefore, for elastic data traffic ($N_f = 1$), the BS queue can be considered as an M/G/1 queue. However, for interactive data traffic ($N_f > 1$), a portion of users whose file transfer is completed, place new service requests at the BS at the end of their thinking durations. The request arrival process at the BS is a combination of the service requests placed by the new users and the existing users returning from the thinking state. Therefore, the request arrival process at the BS may not be a Poisson process, and the service at the BS should be modeled as a G/G/1 queue. All the users entering the thinking state can be modeled by a thinking queue with an exponentially distributed service time and an infinite number of servers to serve any number of incoming users. As the arrivals at the thinking queue depend on the departures of the BS queue, which is a general process, the arrival process of the thinking queue is also a general process. The system can be modeled as two cascaded queues with feedback, as illustrated in Fig. 4.2, where the subscripts $i = 0$, $i = 1$, and $i = 2$ denote the session level parameters of the system, request (call) level parameters of the BS queue, and the request (call) level parameters of the thinking queue, respectively. Symbols $x_i(t)$ and $y_i(t)$ denote the arrival and departure processes, respectively, and λ_i and μ_i denote the mean values of $x_i(t)$ and $y_i(t)$, respectively, where $i \in \{0, 1, 2\}$.

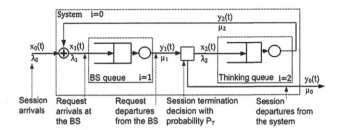

Fig. 4.2 The cascaded queues with feedback [1]

---

[1]A request goes through the waiting queue and the service queue for service completion at the BS, and the BS queue is a combination of both the queues.

The request arrival process to the BS queue $x_1(t)$ is a combination of the Poisson session arrival process $x_0(t)$ and a general departure process $y_2(t)$. Therefore, $x_1(t)$ is a general process. Given the STR distribution, the channel availability statistics, and the service disciplines, we want to characterize $x_1(t)$, which is the first step to analyze the MRT. The thinking queue is a G/M/∞ queue. If the arrival process $x_2(t)$ is Poisson, the it becomes an M/M/∞ queue, and Poisson arrivals at the thinking queue result in Poisson departures with the mean equal to that of the arrival process. This would lead to Poisson arrivals at the BS queue. Since the request arrival process $x_1(t)$ at the BS queue depends on the departure process $y_1(t)$, we need to model $y_1(t)$. The departure processes of the BS queue is complex to analyze with the given service disciplines and the service interruptions (due to the presence of the PUs). As a result, we resort to simulations to study the behavior of the departure process $y_1(t)$ and the arrival process $x_1(t)$ of the BS queue. The behavior characterization is necessary for analyzing the MRT.

4.4 The BS Queue Departure and Arrival Processes

In the following, we use the term distribution to denote the CDF of inter-arrival/departure times.

4.4.1 Inter-Departure Time Distribution (IDTD)

Computer simulations (using Matlab) are carried out to compare the BS queue IDTD with an exponential distribution. In the simulations, all the time durations are measured in discrete-time (in terms of time-slots). Therefore, the channel availability and busy durations are geometrically distributed with mean $1/S_{1,0}$ and $1/S_{0,1}$, respectively. However, for analysis tractability, all the time durations are treated as continuous time durations. Without loss of generality, the data packet size, L_{pk}, is considered as a unit length, and the PDF of the STR, τ, is $f_\tau(x) = f_L(x)$. When comparing an inter-arrival/departure time distribution with a continuous-time distribution using the simulation results, the discrete-time equivalent of the continuous-time distribution is used for the comparison. First, we obtain the inter-departure times of the BS queue with $N_f = 1$ (i.e., elastic data traffic), with a Poisson session arrival process, and $E[\tau] = 200$ time-slots. We carry out the simulations for different channel availability statistics and different tail properties of the STR distributions. For each system parameter set, we complete ten simulation runs, and in each simulation run the BS completes 9,000 service requests.

Figures 4.3–4.6 illustrates the comparison of the inter-departure time distribution and the corresponding geometric distribution for the four service disciplines, respectively. The results demonstrate that, the longer the inter-departure time, the closer its distribution to the geometric distribution. However, the IDTD shows a large deviation from the geometric distribution for the SPTNP service discipline, when the STR is heavytail distributed. For the exponentially distributed STR, the longer the T_{off} and the shorter the T_{on}, we have

- The larger the deviation between the IDTD and the corresponding geometric distribution, when the service disciplines do not allow preemptions;
- The smaller the deviation between the IDTD and the corresponding geometric distribution, when the service disciplines allow preemptions.

Even though the IDTD deviates from the geometric distribution at short inter-departure times, for the tractability of the MRT analysis, we assume that the IDTD of the BS queue is exponentially distributed in the following scenarios: (i) for all four service disciplines, when the STR is exponentially distributed; (ii) For the PS, SRPT, and SPTWP service disciplines, when the STR is heavytail distributed, given that the T_{off} is not very long and T_{on} is not very short in comparison with the mean STR.

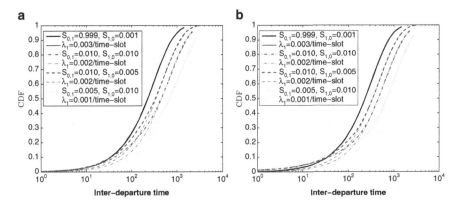

Fig. 4.3 Comparison of IDTD with the geometric distribution for the PS service discipline having (**a**) Exponentially distributed STR, (**b**) Heavytail distributed STR

4.4.2 Inter-Arrival Time Distribution (IATD)

From Fig. 4.2, if the departure process $y_1(t)$ is Poisson, the arrival process $x_2(t)$ is also Poisson with mean $\lambda_2 = (1 - P_T)\mu_1$. The term P_T accounts for the session departures from the system. Therefore, the mean arrival rate at the thinking queue is given by, $\lambda_2 = (1 - P_T)\mu_1 = (1 - P_T)\lambda_1$. Further, the request arrivals at the BS queue is a Poisson process with mean arrival rate $\lambda_1 = \lambda_0 + \mu_2$. With further manipulations, it can be shown that $\lambda_1 = \lambda_0[1 + (1 - P_T)/P_T] = E[N_f]\lambda_0$. Therefore, the arrival process of the BS queue is a Poisson process with mean arrival

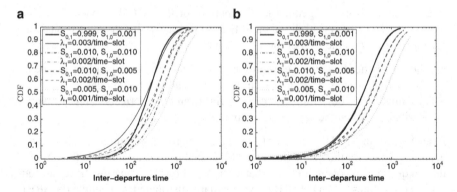

Fig. 4.4 Comparison of IDTD with the geometric distribution for the SRPT service discipline having (**a**) Exponentially distributed STR, (**b**) Heavytail distributed STR

rate $\lambda_1 = E[N_f]\lambda_0$, and the offered traffic load at the BS from the interactive (multi-file) data sessions is equal to $E[N_f]$ times the offered traffic load by single-file data sessions with the same STR distribution. In order to find the closeness of $x_1(t)$ to a Poisson process with mean arrival rate $E[N_f]\lambda_0$, we compare the service request IATD of the BS queue with an exponential distribution with mean $1/E[N_f]\lambda_0$.

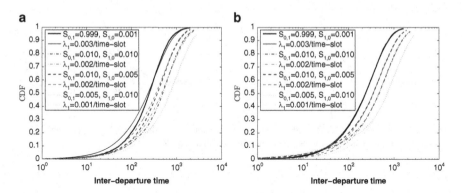

Fig. 4.5 Comparison of IDTD with the geometric distribution for the SPTWP service discipline having (**a**) Exponentially distributed STR, (**b**) Heavytail distributed STR

Computer simulations using Matlab are carried out to compare the service request IATD of the BS queue with the geometric distribution. We obtain the service request inter-arrival times of the BS queue for different $E[N_f]$ values, having a Poisson session arrival process and Weibull distributed STR with $E[\tau] = 200$ time-slots. The STRs are rounded off to the next higher integer value. We carry

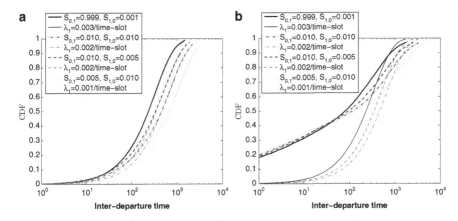

Fig. 4.6 Comparison of IDTD with the geometric distribution for the SPTNP service discipline having (**a**) Exponentially distributed STR, (**b**) Heavytail distributed STR

out the simulations for different channel availability statistics and different tail properties of the STR distribution. For each system parameter set, the BS completes 30,000 service requests.

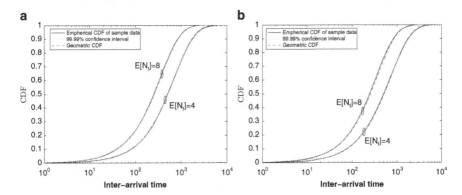

Fig. 4.7 Comparison between the service request IATDs and the geometric distribution for the PS service discipline having (**a**) Exponentially distributed STR, (**b**) Heavytail distributed STR

Figures 4.7–4.10 compare the service request IATD of the BS queue for multi-file data sessions with the geometric distribution having the mean $1/E[N_f]\lambda_0$, $T_{on} = 200$, and $T_{off} = 100$ time-slots. We keep $E[T_T] = 5,000$ time-slots for all the four service disciplines. It can be observed that the service request IATD matches closely with the geometric distribution with mean $1/E[N_f]\lambda_0$, and the match is closer than that between the IDTD and the geometric distribution for all the four service disciplines. From Figs. 4.7–4.10, when T_{on} is longer and T_{off} is shorter

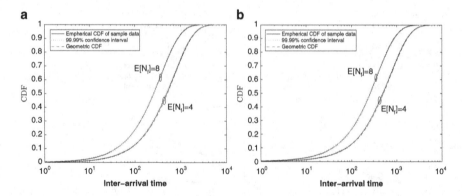

Fig. 4.8 Comparison between the service request IATDs and the geometric distribution for the SRPT service discipline having (**a**) Exponentially distributed STR, (**b**) Heavytail distributed STR

than the mean STR, the geometric distribution stays within the 99.99 % confidence bounds of the observed service request inter-arrival time samples. However, with the SPTNP service discipline, the service request IATD deviates from the geometric distribution, when the mean STR is heavytail distributed. It is due to the large deviation of the IDTD of the BS queue, which is illustrated in Fig. 4.6. Further, it can be shown that, when T_{on} is short and T_{off} is long, the service request IATD slightly deviates from the geometric distribution, and it is observed that, as $E[N_f]$ increases:

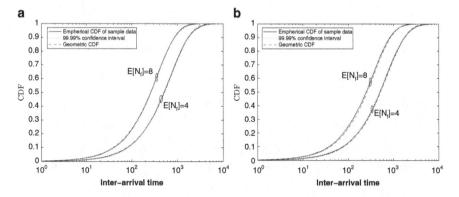

Fig. 4.9 Comparison between the service request IATDs and the geometric distribution for the SPTWP service discipline having (**a**) Exponentially distributed STR, (**b**) Heavytail distributed STR

- There is a close match between service request IATD and the geometric distribution (i.e., in continuous-time, between the IATD and the exponential distribution with mean $1/E[N_f]\lambda_0$);

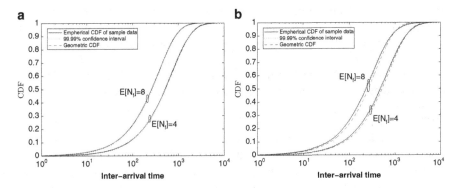

Fig. 4.10 Comparison between the service request IATDs and the geometric distribution for the SPTNP service discipline having (**a**) Exponentially distributed STR, (**b**) Heavytail distributed STR

- The offered traffic load to the BS queue increases, leading to a larger departure rate from the BS queue and a lower chance of having an empty thinking queue.

As the thinking durations are exponentially distributed, the lower the probability of having an empty thinking queue, the closer the IDTD of the thinking queue to the exponential distribution.

Fig. 4.11 Variation of the MRT with $E[N_f]$ for the PS service discipline

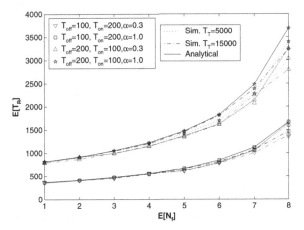

4.5 MRT Evaluation

The simulation result in Sect. 4.4 demonstrate that the service request IATD to matches well with the geometric distribution. As a result, for mathematical analysis of the MRT in continuous-time, the request arrival process at the BS queue is approximated by a Poisson process with mean arrival rate $\lambda_1 E[N_f]\lambda_0$ for a T_{on} longer than and a T_{off} shorter than the mean STR. Denote the mean session arrival rate as λ_0 and the offered traffic load at the BS from the first file of interactive data sessions as the base load, $\sigma_0 = \lambda_0 \cdot L / \pi_1 L_{pk}$, where $\pi_1 = T_{on}/(T_{on} + T_{off})$. Under the Poisson approximation, it is clear that the request arrivals at the BS from a system with multi-file data sessions can be represented by an equivalent system with single-file (elastic) data sessions. Therefore, the MRT for the SPTNP, SPTWP, SRPT, and PS service disciplines can be obtained using the elastic data session MRT analysis in [8], with mean session arrival rate $\lambda = E[N_f]\lambda_0$. Note that, at the steady state, when the arrival process of the thinking queue is Poisson, the mean thinking time has no impact on the mean departure rate. Therefore, the mean thinking time has no impact on the mean arrival rate at the BS queue and the MRT, when the system reaches a steady state. Computer simulations are carried out to evaluate the accuracy of the MRT analysis for the interactive data traffic. The simulation setup is similar to the one used in Sect. 4.4.2 to obtain the inter-arrival/departure times of the BS queue for interactive data traffic ($E[N_f] > 1$). The MRT is evaluated by averaging the results from 20 simulation runs, each run having 18,000 service requests.

Fig. 4.12 Variation of the MRT with $E[N_f]$ for the SRPT service discipline

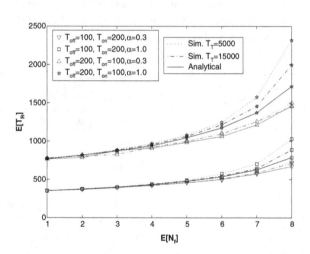

Figures 4.11–4.14 show the relation of the MRT versus $E[N_f]$ obtained from the simulations and the analysis for all the service disciplines, respectively. We keep $E[\tau] = 200$ time-slots and $\sigma_0 = 0.1$ throughout the simulations.

It is observed that the smaller the $E[N_f]$, the closer the match between the simulation and analytical results for (i) for the SPTWP, SRPT, and PS service disciplines with both the STR distributions, and (ii) the SPTNP service discipline when the STR is exponentially distributed. When $E[N_f]$ is large, and the offered traffic load is large, a close match between the MRT simulation and analytical results is expected. This is because, from Figs. 4.7–4.10, both the IDTD and IATD of the BS queue closely follow a geometric distribution. On the other hand, as shown in

Fig. 4.13 Variation of the MRT with $E[N_f]$ for the SPTWP service discipline

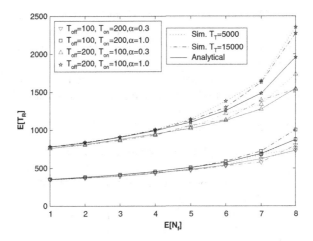

[8], the rate of MRT increment with the offered traffic load at the BS is larger as the traffic load increases. Therefore, when the offered traffic load is large, the sensitivity of the MRT to variations of the offered traffic load is high, resulting in larger MRT variations in response to small variations in the offered traffic load at the BS. As a result, the deviation between the MRT simulation and analytical results increases with the offered traffic load.

Fig. 4.14 Variation of the MRT with $E[N_f]$ for the SPTNP service discipline

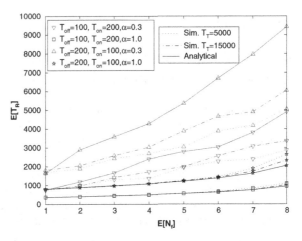

There is a large deviation between the MRT simulation and analytical results for the SPTNP service discipline, when the STR is heavytail distributed. It is due to the slight deviation of the service request IATD from the exponential distribution. Therefore, we can conclude that the MRT under the SPTNP service discipline is more sensitive to the variations of the service request IATD when the STR is heavytail distributed than when it is exponentially distributed. That is, the sensitivity

of the MRT to the variations of the service request IATD depends on the tail
heaviness of the STR distribution. It has been shown that the SPTNP is not a good
choice when the tail of the STR distribution is very heavy [8]. Further, for the other
three service disciplines, the sensitivity of the MRT to the variations of the service
request IATD does not vary much with the tail heaviness of the STR distribution.
Figures 4.11–4.14 demonstrate that, a shorter (mean) thinking time leads to a larger
deviation between the simulation and analytical results for the MRT. When the mean
thinking time decreases:

• The requests leave the thinking queue faster;
• The chances of having an empty thinking queue increase;
• The deviation between the IDTD of the thinking queue and the exponential
  distribution increases; and
• The deviation between the IATD of the BS queue and the exponential distribution
  also increases.

Therefore, the MRT analytical results deviate more from the simulation results.
It is clear from the numerical results that MRT of the interactive data traffic can be
closely approximated by that of the equivalent elastic data traffic, in low and medium
traffic load conditions, in the following scenarios: (i) for all four service disciplines
with an exponentially distributed STR; and (ii) for the PS, SRPT, SPTWP service
disciplines with a heavytail distributed STR.

4.6   Summary

In this chapter, we have investigated the relationship between the interactive and
elastic data traffic flows, in terms of the MRT. Given that the session arrivals follow
a Poisson process, it is shown that the departure process of the BS queue deviates
from the Poisson process when the mean channel available duration is shorter and
the unavailable duration is longer than the mean STR. However, the request arrival
process at the BS can be approximated by the Poisson process under specific channel
availability conditions. The request arrival/departure processes of the BS queue
show a larger deviation from the Poisson process for the SPTNP service discipline
when the mean STR is heavytail distributed. The MRT of multi-file data sessions can
be approximated by that of the single-file (elastic) data sessions in low and medium
equivalent offered traffic load conditions at the BS. Significant deviations between
the analytical and simulation results are observed in a heavy traffic load condition.
Further, it is shown that mean thinking time has no significant impact on the MRT
in low and medium traffic load conditions when the system is at the steady state.
As web browsing is one of the most common interactive data service, the insights
of this study will help to develop packet scheduling schemes and call admission
control algorithms for internet access over CRNs.

References

1. Song, W., Zhuang, W.: Multi-class resource management in a cellular/wlan integrated network. In: Proc. IEEE WCNC, pp. 3070–3075 (2007)
2. Song, W., Zhuang, W.: Resource allocation for conversational, streaming, and interactive services in cellular/wlan interworking. In: Proc. IEEE GLOBECOM, pp. 4785–4789 (2007)
3. Shankaranarayanan, N.K., Jiang, Z., Mishra, P.: Performance of a shared packet wireless network with interactive data users. Mobile Networks and Applications **8**(3), 279–293 (2003)
4. Wang, P., Niyato, D., Jiang, H.: Voice service capacity analysis for cognitive radio networks. IEEE Trans. Wireless Commun. **59**(4), 1779–1790 (2010)
5. Lee, H., Cho, D.H.: Voip capacity analysis in cognitive radio system. IEEE Commun. Lett. **13**(6), 393–395 (2009)
6. Song, W., Zhuang, W.: Multi-service load sharing for resource management in the cellular/wlan integrated network. IEEE Trans. Wireless Commun. **8**(2), 725–735 (2009)
7. Rezaul, K.M., Pakštas, A.: Web traffic analysis based on edf statistics. In: Proc. 7th Annual Post Graduate Symposium on the Convergence of Telecommunications, Networking and Broadcasting (PGNet) (2006)
8. Gunawardena, S., Zhuang, W.: Service response time of elastic data traffic in cognitive radio networks. IEEE J. Select. Areas Commun. **31**(3), 559–570 (2013)

Chapter 5
Conclusions and Future Directions

In this chapter, we summarize the main research results and identify some further works.

5.1 Research Contributions

The objective of this Brief is to study the voice capacity of CRNs and the mean response time of elastic/interactive data traffic over the CRNs with different service processes. As there is an uncertainty in spectrum resource availability, rather than a strict delay requirement, a stochastic delay requirement is considered as the main QoS parameter for voice capacity analysis. The voice capacity is given in terms of the number of single-hop (up-link) voice traffic flows that can be supported by the system. A packet level analysis of the source buffer is carried out to find the packet dropping probability due to violation of the delay bound. The results demonstrate that the silent suppression in the voice traffic (on-off voice traffic) provides approximately twice the capacity than that of the constant-rate voice traffic. An existing DTMC model is modified to analyze the capacity of the slot-ALOHA scheme, and a new DTMC model is developed to analyze the capacity of the round-robin scheme in supporting constant-rate voice traffic over distributed fully-connects CRNs. It is shown that the round-robin scheme performs better than the random allocation and the slot-ALOHA schemes. Further, the impact of the number of packets that can be transmitted in a time-slot (per channel) on the system capacity is larger in the round-robin scheme than in the other two schemes. As the overhead required in implementing the FCFS service discipline is larger than that of implementing round-robin scheme, the round-robin scheme can be considered as an alternative in a centralized network. Given the mean channel availability, the system capacity decreases with the mean channel unavailable duration. Therefore, it is clear that the mean channel unavailable duration has a significant impact on the system capacity. For the simplicity in the capacity analysis, effects of the sensing

S. Gunawardena and W. Zhuang, *Modeling and Analysis of Voice and Data in Cognitive Radio Networks*, SpringerBriefs in Computer Science,
DOI 10.1007/978-3-319-04645-7_5, © The Author(s) 2014

errors has not been considered in our analysis. However, we have discussed the possibility of extending the DTMC models to incorporate the sensing errors into the capacity analysis. The analytical models used for the constant-rate voice capacity based on the stochastic delay requirement can be used to analyze the capacity of any type of constant-rate traffic flow. Further, our capacity analysis can be used as benchmarks in the performance analysis phase of the development of new channel access schemes to support voice traffic over CRNs. As the capacity analysis is limited to homogeneous voice traffic flows, two CAC algorithms are developed to support constant-rat voice traffic with different delay requirements (different delay bound and maximum delay bound violation probability) over non-fully-connected slot-ALOHA CRNs based on an empirical relationship between the successful transmission probability and the delay bound violation probability. It is shown that the longer the delay bound, the larger the system capacity. In other words, the lower the required service quality, the higher the system capacity. A low quality service can be priced at a lower rate than a high quality service to increase the user satisfaction

This brief presents a study on the mean response time of elastic data traffic via a session level analysis. Three service disciplines, namely, shortest processor time without preemption, shortest processor time with preemption, and shortest remaining processing time are studied in comparison with the processor sharing service discipline. The works on non-cognitive networks only consider the mean channel availability (the long term channel availability) in the mean response time analysis. However, we have shown that the variation of the mean channel available/non-available durations have a significant impact on the mean response time even when the mean channel availability is constant. Therefore, the mean response time evaluations based on the mean channel availability are not suitable in the context of the CRNs. The SRPT service discipline outperforms the other three service disciplines, and the preemption reduces the mean response time for heavytail distributed STRs. However, if the original STR is compared for preemptions rather than remaining STR, the advantage of the preemptions disappears with decreasing tail heaviness of the STR distribution. Therefore, when the STR distribution is concentrated around its mean, preemption gives a negative impact on the mean response time, and the SPTWP gives longer mean response times than the SPTNP. Even though the equal bandwidth sharing scheme in the PS outperforms the SPTNP when the tail of the STR is heavy, the lighter the tail of the STR distribution, the shorter the mean response times of the SPTNP in comparison with that of the PS. Further, the heavier the traffic load, the larger the rate of increment of the mean response time for the PS than that for the SPTNP. However, when the tail of the STR is heavier, it is shown that the mean response times corresponding to the SPTNP are extremely longer than that corresponding to the other three service disciplines. Therefore, the SPTNP service discipline is not a good choice when the tail of the STR distribution is very heavy, and it is a better choice over the PS when the tail of the STR distribution is light. The behavior of the BS of a CRN is analogues to a service station with random breakdowns, in a machine repair problem in the operations and research studies. Therefore, this analysis can be used in machine repair problems with service station subjected to breakdown, considering the same service disciplines at the service station.

We have studied the interactive data service over CRNs by modeling the system as two cascaded queues with feedback. It is shown that the request arrival process at the BS for interactive data traffic is close to the Poisson process, given Poisson session arrivals and longer (shorter) mean channel available (unavailable) durations in comparison with the mean STR. A relationship between the mean response times of elastic data traffic and the interactive data traffic is obtained under the Poisson approximation of the request arrivals at the BS. As an interactive data traffic session represents the behavior of a Web traffic user, the obtained relationship between the mean response times of interactive and elastic data traffic can be used for QoS provisioning in Web traffic over CRNs.

5.2 Future Directions

In Brief, we consider a time-slotted primary network in which the PU activities are perfectly synchronized with the time slots. Further, we assume the capability of the SUs to synchronize with the time-slots and to carry-out perfect channel sensing. However, the sensing errors are inevitable in the practical networking scenarios. If the presence of a PU is not detected at the spectrum sensing stage (by SUs), it is denoted as a missed detection error, and it can be caused by channel fading. The missed detections lead the SUs to transmit simultaneously with the PUs, which may cause harmful interference to the PUs. As a fundamental requirement in CRNs, it is required to limit the interference to the PUs below a predefined threshold by minimizing the missed detection errors. The interference to the PUs can be quantified by the probability, P_{md}, of missed detection and the duration of interference. In the network under consideration, the duration of interference is the time-slot duration,[1] which is a fixed quantity. Therefore, the interference to the PUs is characterized by P_{md}, which should be controlled below a certain threshold, P_{md}^*, to satisfy the condition on interference with the PUs. The longer the sensing duration, the more accurate the sensing decision and the shorter the time available for information packet transmission [1]. There is a trade-off between the sensing accuracy and the spectrum time available for the secondary network. In the voice and data networks under consideration, the shorter the time available for information packet transmission, the smaller the n_{pk} and L_{pk}. It is shown in Sect. 2.6 that the voice capacity monotonically increases with n_{pk}. Further, in Sect. 3.5 it is shown

---

[1]When the primary network is time-slotted, as the worst case scenario, whole information frame is assumed to be lost due an interference caused by the secondary network. However, the actual interference duration depends on how the information packets are organized and transmitted within the time-slot (by a PU).

that the longer the STR, the longer the mean response time. However, the longer the L_{pk}, the shorter the STR. Therefore, the longer the L_{pk}, the shorter the mean response time, and the larger the n_{pk}, the larger the voice capacity. Further, the smaller the probability, P_{fa}, of false alarm (the accurate the sensing decision), the larger the amount of spectrum opportunities available for the secondary network, leading to larger the voice capacities and shorter the mean response times. For the optimum performance of the systems under consideration, the voice capacity should be maximized and the mean response time of elastic/interactive data traffic should be minimized, while satisfying the constraint $P_{md} \leq P_{md}^{*}$.

In a CRN network with a common channel (known as common control channel) to share the sensing decisions, the sensing decisions of multiple SUs can be combined to obtained the final sensing decision. This procedure is known as cooperative spectrum sensing [2], which results in a better sensing accuracy. It is an interesting research area in the context of CRNs. Even with the application of cooperative spectrum sensing, missed detections (interference with the PUs) are inevitable. In the centralized CRNs under consideration, cooperative spectrum sensing can be accomplished via a control channel, by incorporating sensing decisions of multiple SUs at the BS to make the final decision. The BS can transmit the final sensing decision back to the SUs. In this way, the consistency in spectrum availability information can be achieved. In the presence of cooperative spectrum sensing, the missed detections and the false alarms probabilities are reduced. However, cooperative spectrum sensing can only be accomplished at the expense of available spectrum time for the secondary network, which has a negative impact on n_{pk} and L_{pk}. There is a trade-off between the gain in the spectrum availability by reducing the probability of false alarms and the spectrum time lost due to the cooperation. For the simplicity of our analysis, we did not incorporate the sensing errors and the interference to the primary users in our analysis. However, as the spectrum sensing errors are inevitable, and it is worthwhile to study their effects on the voice capacity and the mean response time of elastic/interactive data traffic.

So far, we have discussed the scenarios with a time-slotted primary network (synchronous primary network). However, it is important to study the CRNs operating over asynchronous primary networks. Even though the primary network is not synchronized, the secondary network can operate in a time-slotted manner with the accomplishment of periodic channel sensing. In this type of CRNs, in addition to the missed detections (occur at the sensing periods), the presence of a PU in between the finishing and stating time instances of two adjacent sensing periods, respectively, can cause interference to the primary network. Therefore, the interference to the primary network is larger than that when the primary network is synchronized (time-slotted). Due to the asynchronous nature of the primary network, the interference duration is also different from that when the primary network is synchronized. The interference duration due to a missed detection depends on the probability distribution of the duration that a PU occupies the channel, whereas the interference duration due to the presence of a PU in between the finishing and the starting time instances, receptively, in adjacent time-slots depends on the probability distribution of the duration that a PU is idle. Given the corresponding probability distributions, the probability of interference to the primary network can be analyzed. In order to optimize the network performance, the sensing period and the sensing duration need to be fine-tuned such that the voice capacity is maximized (or the mean response time of elastic/interactive data is minimized), while the interference with the primary network is kept under control. A comprehensive study on the synchronous CRNs operating over asynchronous primary networks will be a significant contribution to the CRN research.

References

1. Pei, Y., Hoang, A.T., Liang, Y.C.: Sensing throughput trade-off in cognitive radio networks: how frequently should spectrum sensing be carried out? In: Proc. IEEE PIMRC, pp. 5330–5335 (2007)
2. Mallik, R.K., Letaief, K.: Optimization of cooperative spectrum sensing with energy detection in cognitive radio networks. IEEE Trans. Wireless Commun. 8(12), 5761–5766 (2009)